D0768248

HOW TO TEACH

Respiration is not breathing!

Secondary
Science

CATRIN GREEN
EDITED BY PHIL BEADLE

 Independent Thinking Press

CALGARY PUBLIC LIBRARY

FEB   2017

First published by

Independent Thinking Press
Crown Buildings, Bancyfelin, Carmarthen, Wales, SA33 5ND, UK
www.independentthinkingpress.com

Independent Thinking Press is an imprint of Crown House Publishing Ltd.

© Catrin Green 2016

The right of Catrin Green to be identified as the author of this work has been asserted by
her in accordance with the Copyright, Designs and Patents Act 1988.

Illustrations © Les Evans 2016 (except pages 20 and 28, illustrations by Jonathan Tran)

The right of Les Evans to be identified as the illustrator of this work has been asserted by
him in accordance with the Copyright, Designs and Patents Act 1988.

First published 2016.

All rights reserved. Except as permitted under current legislation no part of this work may be photocopied,
stored in a retrieval system, published, performed in public, adapted, broadcast, transmitted, recorded or
reproduced in any form or by any means, without the prior permission of the copyright owners. Enquiries
should be addressed to Independent Thinking Press.

Independent Thinking Press has no responsibility for the persistence or accuracy of URLs for external or third-
party websites referred to in this publication, and does not guarantee that any content on such websites is, or
will remain, accurate or appropriate.

Edited by Phil Beadle

British Library Cataloguing-in-Publication Data
A catalogue entry for this book is available
from the British Library.

Print ISBN 978-1-78135-241-0
Mobi ISBN 978-1-78135-258-8
ePub ISBN 978-1-78135-259-5
ePDF ISBN 978-1-78135-260-1

Printed and bound in the UK by
Bell & Bain Ltd, Thornliebank, Glasgow

ACKNOWLEDGEMENTS

When I first started teaching I was very lucky to meet Phil Beadle, who both inspired me and challenged me to become a better teacher. I am very glad and thankful to have worked alongside him again with this book. My thanks must also go to all of the staff at Crown House Publishing, particularly to Emma Tuck and Caroline Lenton for their help throughout the writing of this book.

Thank you to all of my family, in particular Jon, for always encouraging me to get outside of my comfort zone. This book is dedicated to my son Ioan, who was with me throughout its writing.

CALGARY PUBLIC LIBRARY

FEB   2017

FOREWORD BY PHIL BEADLE

'Open your books at page 34. Do exercises 1 through to 17 and shut up while you're doing them!'

I've sat in on some quite poor science lessons in my time (two in a row, in fact, where one of the key words was 'bucket'). The 'not really trying very hard' version of these lessons always seemed to be orientated around a stained textbook. The kids were guided through a reasonable enough practical activity but were *always* forced to follow this up with a series of drab comprehension questions that the students answered entirely perfunctorily, all the time studiously ignoring the teacher's vain and frustrated wish that they write in full sentences. We didn't learn much. But the time passed. And we were, all of us, one day nearer the day we got out of this dump.

It was a shame that this brilliant subject was marginalised as the blind, impotent witch in the triumvirate of important things to know (maths, English and the other one). I remember thinking a decade ago that it could be so much more than it was at that time.

Enter Teach First. Enter Catrin Green.

The first time I sat in a science lesson and thought, 'Oh, *this* is how it should be done,' was in a knackered lab in an academy in the outer reaches of Croydon. Here, a young woman, the author of this book, was taking risks, expecting the kids to understand difficult things, playing with the form. It was relatively early on in her first term as a teacher and she was already really, really good. So good, in fact, that they offered her the head of science post before the end of the second term. (I believe she said no as she wanted to focus on being as good a teacher as she could be.)

But it was not just the lesson, profoundly impressive as it was, or the kids' achievement and enjoyment, which was tangible, that impressed. It was in the feedback afterwards, in which Catrin, fizzing with intelligence, started questioning some of the Ofsted tropes and some of the thoughtless givens of 'fashionable' pedagogy at the time. I recall thinking that if Catrin Green was in any way representative of the kind of teacher Teach First were bringing into the profession, then I would have to put my cynicism about it away.

Science teaching has, I feel, improved a lot since I spent much of my life sitting at the back of other people's lessons ticking and crossing silly boxes. Things move on. And Catrin is no longer a talented ingénue sparking with potential. The book you have in your hands is written by a seasoned and talented teacher with nearly 10 years' experience of getting kids in Croydon to learn science, to love science and to love learning science. There will be bits that you disagree with, and there might be points at which you throw the book across the room (Catrin's approach is – dare I say – quite 'progressive' at points); but what you have here are the thoughts and ideas of an excellent practitioner who always finds a better way of doing things than, 'Open your books at page 34. Do exercises 1 through to 17 and ...'

CONTENTS

INTRODUCTION

In some ways, science teaching hasn't moved on that much since you were at school. The formation of ionic bonds, identifying the organelles of the cell and the difference between voltage and current are still mainstays in the vast catalogue of content we have to teach. However, just as scientific understanding itself has moved on, so has the way we teach it.

Nowadays, science labs are rarely rows of benched students working through questions from a textbook or copying notes from the board. Science teachers have moved on from the 'This is the truth – now get on with it' mentality I remember from the lessons I was subjected to at school towards a focus on lively and engaging ways of teaching the key concepts. This shift, though, can pose a challenge to teachers as exciting student-centric lessons can be time consuming to plan and, if this planning hasn't been properly undertaken, they can result in students learning little or nothing. This book aims to show you how you might plan memorable (and, yes, even fun) lessons in a time efficient way that, most importantly, has learning at the centre.

Science is more important and more high profile than ever before: 92% of firms across all sectors require staff with science, technology, engineering and mathematics (STEM) backgrounds and most struggle to recruit the right people.[1] If we are going to plug the gap in numbers, science teachers need to teach cutting edge content and provide new experiences for our students in order to pass on our passion. Many adults have gone on to be totally enthralled by popular science after they've left school, but if you asked them what their memory of school science was, they would have reported

---

1   See http://www.policyexchange.org.uk/publications/category/item/
     science-fiction-uncovering-the-real-level-of-science-skills-at-school-and-university.

1

ambivalence to the lessons at best and active dislike at worst. It is incumbent on practitioners to ask why this is, what we might do about it and how we might change things so the science curriculum comes alive.

Additionally, people can so often be easy victims to the dupery of 'bad science' because they don't have a good enough grasp of scientific knowledge to see through the guff. A student in your classroom could go on to work in the media or in the next area of pseudo-science (we will not be giving any time here to homeopathy etc.). As such, it is our duty as science teachers to ensure that young people are able to make informed decisions in the future – and we do not end up producing the next Gillian McKeith or 'science' correspondent for the *Daily Mail*.

Most science teachers have a degree in the subject and ought to be pioneers in using techniques founded in the theory of science. Ongoing research from cognitive psychology should be used to inform how we teach, so this book draws on a number of ideas sourced from Hattie's well-known meta-analysis.[2] Whilst teaching is never going to be a physical science, and whilst acknowledging that what works in an inner city school may not work in a rural grammar, research can show us new ways to think about our practice and look for ways to become better teachers. What unifies all the ideas in this book are that they are designed to engage your students and to make sure they learn; you should be able to use them to support students regardless of your school setting.

I have always had an active classroom – we regularly sing, dance and play games to learn new content – but we also believe in challenge. These two ideas are mutually exclusive only to the myopic or the bigoted. It is possible that some of these ideas, at first attempt, may result in a perceived failure, but this can often be merely an 'implementation dip' and it is important that

---

2   J. Hattie, *Visible Learning: A Synthesis of Over 800 Meta-Analyses in Education* (Abingdon: Routledge, 2008).

you keep trying with them. We can't teach our students to learn from and embrace their mistakes if we are not happy to do the same.

Successful science teaching is all in the balance: the balance between knowledge and skills; the balance between student engagement and hard work; the balance between teaching incredible lessons all day, every day, and managing to maintain any semblance of sanity. This book is a guide to finding the middle ground on all these issues. It will give you plenty of hands-on ideas about how to make the teaching of scientific ideas memorable, without you planning lessons until midnight and arriving at your class with bags under your eyes and a look of horror on your face, rather than the big smile that is a teacher essential.

Danielle McNamara observes: 'There is an overwhelming assumption in our educational system that the most important thing to deliver to our students is content. Teachers assume that when they have covered something in a course that it should be absorbed by the student.'[3] Science teachers *do* have to deliver a lot of information to students, and we need to find interesting ways of making that knowledge mean something to young people so they can remember it and create their own understandings. The number one issue all science teachers talk about is how difficult it is to pass on all the knowledge in the curriculum whilst also extending and developing scientific skill. Any glimpse at a GCSE science specification will see the word 'recall' littered over every page, but if all we do is share information then our students won't develop real understanding of it. And just because we have lots of knowledge to communicate, this doesn't mean that we should always fall back on resorting to chalk and talk as a default setting. This has its place, of course, as there are times when explicit instruction is necessary for clear understanding (such as when introducing the concepts of reacting masses

3   D. S. McNamara, 'Strategies to Learn and Learn: Overcoming Learning by Consumption', *Medical Education* 44(4) (2010): 340–346.

or half-life). However, I find that using a variety of teaching tools is the best way to achieve the balance we are after.

Even if you are able to give your students the most memorable experiences possible, this may well not be enough to ensure that they will be able to commit such information to long term memory. You will also need to establish that they are not labouring under misconceptions, you will have to give them chances to really think and you will need to provide opportunity after opportunity for them to practise what they have learnt. The following chapters outline how to achieve good learning in science classrooms.

Chapter 1

THE IMPORTANCE OF THE BIG PICTURE

'BUT WHAT HAVE PARTICLES GOT TO DO WITH REAL LIFE?!'

The science curriculum can appear as a 'catalogue' of discrete ideas, lacking coherence or relevance. There is an over-emphasis on content which is often taught in isolation from the kinds of contexts which would provide essential relevance and meaning.

Robin Millar and Jonathan Osborne[1]

To engage students in learning we provide them with a peg on which to hook their new learning – usually background knowledge from day-to-day life, from the previous lesson or from a prior topic. What a student already knows about a subject has a much greater impact on achievement than both the interests of the student and the skill of the teacher,[2] and whilst it is not possible for a teacher to fully influence a student's background knowledge, one of the most important factors in its acquisition in the first place is the

---

1   R. Millar and J. F. Osborne, *Beyond 2000: Science Education for the Future* (London: King's College London, 1998), p. 3.

2   See K. Halikari, N. Katajavuori and S. Lindblom-Ylänne, 'The Relevance of Prior Knowledge in Learning and Instructional Design', *American Journal of Pharmaceutical Education* 42(7) (2008): 712–720. Daniel T. Willingham's *Why Don't Students Like School? A Cognitive Scientist Answers Questions About How the Mind Works and What It Means for the Classroom* (San Francisco, CA: Jossey-Bass, 2010) provides an easy-to-read and interesting overview of the impact of background knowledge on learners.

number of opportunities that we provide students with to understand the content and how we find ways of linking the science to real life.

TELLING THE STORY

Although we would probably all agree that science is an awe inspiring subject, scientists (science teachers included) need to become much better at communicating this awe to those who are not yet scientists. All students are curious – no matter how apathetic they may initially appear – and the trick is to find the hook with which to engage them. One of the best ways of doing this can be to bring the science to life with a story. This isn't a new idea, of course, as linking knowledge to a story and creating a narrative has long been a key way of developing knowledge and learning. Not only does a story allow students to interconnect ideas they might already have about science, but it also allows them to place their knowledge in a wider context.

Storytelling is perceived as central to learning in English or history lessons, and whilst it might seem slightly more difficult (or even counterintuitive) in science, it's actually pretty easy. Think about any science documentary you've ever seen on TV – this is always the method they use to introduce new topics. Not only is it engaging but it also provides an anchor for the new knowledge. Additionally, provided they are not too bogged down in unnecessary detail, stories are easy to remember: psychologists believe they are treated differently in the memory to any other kind of material.[3] Students often struggle with the fine detail, so if we launch straight into the nitty-gritty of any topic they will quickly ask, 'Why are we doing/learning this?' Showing them the bigger picture and leading them towards being interested

---

3   This article provides a good review of the research: D. T. Willingham, 'Ask the Cognitive Scientist', *American Educator* (summer 2004). Available at: http://www.aft.org/periodical/american-educator/summer-2004/ask-cognitive-scientist.

is a good start point. Then the students will start to ask questions: 'Yes, but how?' or 'What next?' You might wonder what this has to do with learning, but studies have shown that teaching students the cognitive strategy of asking questions results in significant gains in comprehension.[4] A random list of 10 numbers with no connections is difficult to learn, but link the numbers to things in your own life and suddenly it's not quite so hard after all.[5] Therefore, a student who is simply taught what electrons, protons and neutrons are (without any surrounding context) may struggle to understand their relevance, but a student who has been taught about how our understanding of the basic building blocks of life has evolved has a connection between the different ideas and gets why they are studying it and how it is relevant to them.

HOW CAN I GET IT RIGHT FROM THE START?

A simple idea to start introducing stories into your teaching is to think pretty hard about the title of your lesson. So, a lesson that might otherwise be called 'The atom', could be 'What are we really made of?' or 'What is the smallest thing we know about?' Bill Bryson's *A Short History of Nearly Everything* is a fantastic place to start if you need some inspiration for stories.[6]

---

4   See B. Rosenshine, C. Meister and S. Chapman, 'Teaching Students to Generate Questions: A Review of the Intervention Studies', *Review of Educational Research* 66 (1996): 181–221.

5   Read Ed Cooke's *Remember, Remember: Learn the Stuff You Thought You Never Could* (London: Viking, 2008) for ways to memorise lots (like the whole of the periodic table – it would really impress your students. Disclaimer: I cannot do this!).

6   B. Bryson, *A Short History of Nearly Everything* (London: Black Swan, 2003).

Choose the most interesting sections of the story, litter it with some little side facts that spark interest and employ photos and diagrams to illustrate.[7]

Here are some examples of how to storify science for students.

Biology: 'Are we really 97% the same as chimpanzees?'

Evolution can be a tricky topic to teach, mainly because students walk into your classroom with existing ideas and misconceptions about what it is. Starting with evolution as a story is a really good way in: I begin this subject with the story of Darwin during his time at Cambridge. I tell them that he was so curious by nature that he used to eat owls and hawks, and that one of the results of this curiosity was his trip to the Galapagos Islands. At this point, it's important to note that a story doesn't have to be told just by you but can be investigated through group work and projects. For instance, allocate each group of students a different segment of Darwin's story: one group could be given information on his journey, another what he discovered on Galapagos with regards to finches, another how he tried to convince the public and so on. Get them to present these in date order.

---

7   Check out this article by Martin Robinson for tips on how to tell a story like a pro: 'Classroom Practice: Don't Just Talk at Them, Spin a Ripping Yarn', *TES* (14 February 2014). Available at: https://www.tes.com/article.aspx?storyCode=6403314.

Biology: 'Why do I have to have that injection?'

For an account of Edward Jenner, who pioneered the smallpox vaccine, start with a picture of someone suffering from the later stages of smallpox (warn the students first!) and it is likely you will inspire the awe that you are seeking. Continue by painting the story of how deadly smallpox was (some estimates suggest that between 300–500 million people have died of the disease – a higher fatality rate than both world wars combined[8]). Next – and this is a great opportunity to show how medical advances can come from thinking outside of the box – explain how Jenner went from hearing that milkmaids who had contracted cowpox almost never contracted smallpox and that this led him to try inoculating subjects with cowpox before exposing them to smallpox to test his hypothesis. Continue this into the present day by observing that smallpox could be used in bio-terrorism and why it is essential that students do not listen to any of the scare stories regarding vaccinations (after all, your students will be parents one day).

Biology: 'Why do 100,000 people die of cholera every year but I've never heard of it?'

John Snow's (no, not the guy from *Game of Thrones*) discovery of cholera in 1849 was incredible given that he couldn't see bacteria. Cholera was originally thought to be airborne until Snow looked into a particular case in Soho, London. Show students the same maps that he looked at of where people had been infected and prompt them with some questions. What patterns can they see? Would this pattern support the theory that the infection was

---

8   D. Perlin and A. Cohen, *The Complete Idiot's Guide to Dangerous Diseases and Epidemics* (Indianapolis, IN: Alpha, 2002).

airborne? Why/why not? How else could it have been spread (it's best if students already have a bit of background on communicable diseases first)? Once students have suggested that it could be spread through water, show them a second map with locations of the various water wells and see if they can identify the one on Broad (now Broadwick) Street as the source of the infection. Follow this up with similar exercises into recent epidemics (e.g. the Zika virus or Ebola).

Biology: 'How can babies have three parents?'

This is a great way to introduce a Key Stage 4 genetics topic. Teach the role of the mitochondria through the process and ethics of allowing three person babies.[9] Start the lesson by providing pairs of students with any news clippings you can find about this type of a story (differentiate your material here – some pairs will be able to understand *The Guardian* or *The New Scientist*; others might better engage with a clip from the BBC website). Next, split the pairs so that one student has to oppose the idea whilst the other one agrees. Give them time to prepare before getting them to debate in a 'debating ring' (see Chapter 3).

---

9   Mitochondria are organelles found in eukaryotic cells which contain a small amount of DNA which are responsible for respiration. (This is due to their interesting history: they so closely resemble bacteria that one theory suggests they were formed through the symbiosis of a eukaryotic and prokaryotic cell, hence they contain the DNA which originated from the bacteria.) Mitochondria are normally inherited from the mother (so you are genetically ever so slightly more similar to your mother than your father), but if the mother carries a genetic disease in her mitochondrial DNA, it is possible to have this DNA donated by a third party through a modified version of in vitro fertilisation (IVF). This results in a fertilised egg cell formed from the DNA of three 'parents'.

Biology: 'Why am I like my parents?'

The impact that the discovery of DNA has had on our understanding of inheritance is best started with a clip from *Jurassic Park* (the original one, obviously – there's nothing wrong with showing your age). The film can be used to explain how the park brought dinosaurs back from extinction. Use a think–pair–share activity (see Chapter 2) to ask students if they think this would ever be possible and then get them to brainstorm what they already know about DNA. This is important because aspects of this topic are part of everyday life, so you shouldn't assume the students have little background knowledge. A good activity here is to produce a student timeline at the front of the classroom. Start with a student Charles Darwin at one end of the room (get them to make a Darwin sign to hold up) and ask them to explain what Darwin told us about evolution. The next student along from Darwin should act as Gregor Mendel and should explain his pea experiments (have some YouTube clips up your sleeve to remind the students if they are rusty). Now you need to skip to the 1950s, so leave a largish gap and choose four students to be the pioneers of DNA, using them to play the roles of Rosalind Franklin, Maurice Wilkins, James Watson and Francis Crick. (As an aside, the lack of a Nobel prize for Franklin is a great discussion point to engage the girls.) It's worth finishing with a task regarding environment and genetics which will allow the students to conclude just how much (unfortunately) they are like their parents.

Chemistry: 'How have we got such a range of materials?'

Tell the story of the turning points in chemistry, starting with Aristotle believing that the only four elements were fire, earth, air and water through to the modern day and new wonder materials such as graphene. This is a good way to start teaching elements and compounds before introducing students to the periodic table. Hand out some Lego and ask the students to make simple structures (e.g. house, truck, skyscraper). Then ask them why it is possible that a number of different objects can be made from the same building blocks. Use this analogy to illustrate the difference between elements and compounds (and for those who grasp it quickly, molecules). Then you can finish the lesson by looking into how close we are getting to some of the technologies in *Iron Man*.

Chemistry: 'Why is there always a periodic table in my homework planner?'

This is the story of Dmitri Ivanovich Mendeleev. As the students enter the room, ask them to find the periodic table in their planner and pose the title of the lesson to them. Use a snowball activity to find out what students think (see Chapter 2). Provide them with element cards identifying the key properties (most schools will have a set of these or they can easily be found on the Internet); if you can manage it, a sample of some of the elements is useful as well. Ask the students in pairs to sort them in any way they wish, as long as they can justify why they have grouped them together (e.g. they all have low melting points). Tell the students that the activity they've just completed is the essence of what Mendeleev did to come up with the periodic table and, at this point, go through a timeline activity (like the Darwin example above or get the students to use sequencing cards). For the more able, pose some challenging questions to help them think further (e.g. Why is hydrogen so lonely in the periodic table? Why are some periods cast out on their own?). Once the serious work is complete there are a lot of sing-a-long videos about the periodic table that can be found on YouTube that no student is ever too old or too cool for. Finish the lesson by asking the students the same starting question and perhaps get them to provide an exit pass explaining what they now think the answer is.[10]

Chemistry: 'What will my life be like in 2050?'

This is an opportunity to talk about global warming from the Industrial Revolution onwards. Begin the lesson with some photos on the students' desks to act as a stimulus for discussion (e.g. driverless cars, mock-ups of London flooded by rising sea levels, a nuclear power station, robots) and initially ask them to answer the question in the lesson title. Get them to think about what will have the biggest impact on their lives in 2050 (hopefully the pictures will help with the answers) and steer them towards the devastating impact that rising global temperatures might have on their lives. (The trailer for *An Inconvenient Truth* (2006) is a powerful start here.) Put students into groups and give each group a different stage of the timeline: the Industrial Revolution, the rise of nuclear power, new green technologies and possible future technologies. Give them time to research their area and prepare for feeding back to the class. Get the students to present their findings in date order. Complete the lesson(s) by asking the students to reflect on what may happen by 2050 if some new technologies to produce energy are not up and running in time.

---

10 An exit pass is an elaborate name for a sticky note on which the students write their name and answer, which they must give to you before they leave the classroom – it works best before break and lunch!

Chemistry: 'Why do people hate chemicals?'

People tend to try to avoid chemicals, whether in cleaning products or hair dyes, because they fear the repercussions (despite the fact that everything in the world is actually made from chemicals). We can start dispelling this myth when starting the topic of compounds by looking at how this thinking came about – from the use of lead based products to the assumed danger of E numbers. Start by asking the students if chemicals are safe and encourage them to give some examples of 'unsafe' chemicals (you'll get an odd selection here from CFCs and oil to arsenic and kryptonite!). Next, go through the definition of what a chemical is and ask them to re-evaluate their initial answers. Give groups of students different examples of chemicals that are perceived to be dangerous (or can be if used in the wrong way).

Some ideas of stories you might want to give your groups include:

- Images of Elizabeth I and old adverts for lead based make-up.

- Newspaper clippings on the effect of E102 and E110 on children.

- The impact of thalidomide in the 1950s.

- News articles on the presence of arsenic in many day-to-day foods (e.g. rice, cereals, fruit). Are these foods dangerous?

- Should we have fluoride added to our water? Give details on the debate.

- A newspaper article on the terrifying use of a chemical in our chips/ carpet (whatever's been in the news recently – e.g. acrylamide in baby food).

- Provide students with everyday objects that contain 'chemicals' (e.g. shampoo, window cleaner, paracetamol, plastic lunch box). Get them to research any negatives about them.

Ask the students to share what they have learnt and then have a debate about which of the examples are of genuine concern and which are potentially scaremongering.

Physics: 'Is the Earth the centre of the universe?'

A mere 500 years ago the common belief was that the sun and other planets revolved around the Earth. This provides us with a story that we can use to explain how transitory science can be. Start with Aristotle and his arguments for the geocentric model: the Earth must be still otherwise objects would fly off the surface, and if the Earth was moving why do birds not fly off into space? With some groups you could also discuss the idea that no parallax effect is seen with stars. Pose these questions and see if your class can refute them using the post-Newtonian understanding of motion. Move on to explain how the Church used Aristotle's model – replacing the 'prime mover' with God – and the effect this had later on the acceptance of the heliocentric model. Then along comes Copernicus who spent an age watching planets through a telescope and realised that it was impossible to explain what he saw with the geocentric model. He died a sad man – no one believed his ideas. Then get the students to look at the evidence for the heliocentric model.[11]

---

11 The arguments are all very nicely rounded up in this article: R. Allain, 'How Do We Know the Earth Orbits the Sun?', *Wired* (14 April 2014). Available at: http://www.wired.com/2014/04/how-do-we-know-the-earth-orbits-the-sun/.

Physics: 'How did Becquerel manage to burn himself?'

This is an opportunity to introduce radiation as something that you can't see and to examine the implications this had for the first scientists studying it. Start with a clip of Blinky the Fish in *The Simpsons*: why does the poor little guy have three eyes? Most students will already have some understanding of how dangerous radiation can be, and this can act as a springboard into a discussion about how this was not always the case. Tell the story about Henri Becquerel and how he kept radiation in his pocket (some radium salt in a sealed glass tube), only realising that it could be potentially dangerous once he had got some rather nasty radiation burns (he said of radium, 'I love it, but I owe it a grudge').[12] The story can then move on to Marie Curie and how radiation eventually led to her death from cancer. You can even show the students adverts for radium based beauty products and laugh pitiably at the people who used to use them.

Physics: 'Why is Pripyat a deserted town?'

The story of Chernobyl has much to teach students about nuclear power and the importance of making sure that science is safe. It also features in one of the scenes from *Call of Duty* (which, so I've heard, is a popular computer game) and I often show pictures of this to start the topic. Some emotive pictures of direly ill victims of the explosion work well here before moving on to an explanation of what happened (including the fact that extra radiation was detected as far away as the UK). You can find many first-hand accounts

---

12   See R. F. Mould, 'Pierre Curie, 1859–1906', *Current Oncology* 14(2) (2007) : 74–82. Available at: http://www.ncbi.nlm.nih.gov/pmc/articles/PMC1891197/.

online to read to the students (or employ an extrovert student to do the same), including some very moving ones from the staff who had to clear up the mess.[13] Make sure you finish with present day pictures of the town,[14] or some of the recent research which has found that the radiation might be less damaging than previously thought.[15]

Physics: 'What is the Large Hadron Collider trying to do? And will it cause the end of the world?'

The number of students who asked me this question the first time the Large Hadron Collider (LHC) was used was quite astonishing. News items like this are perfect for one-off lessons to help your students understand the reason why they are studying science in a topical way. When 'big science' related news events occur, you can plan lessons or starters to go into the facts behind the story and then differentiate the explanation and tasks for different ability groups. Even with a GCSE class, there will be time (if only for 10–15 minutes), and it is well worth doing. When the LHC went live in 2008, I started the lesson with the headline from that day's *Metro* regarding the end of the world, followed by a discussion on the science behind the story and backed up with a good YouTube clip (never underestimate how well a video can support your explanations, no matter how good you are).

---

13 See for example: M. Block, '"Voices From Chernobyl": Survivors' Stories', *NPR* (21 April 2006). Available at: http://www.npr.org/2006/04/21/5355810/voices-of-chernobyl-survivors-stories.

14 There are some haunting ones here: http://www.telegraph.co.uk/news/picturegalleries/worldnews/9128776/Photographs-of-Chernobyl-and-the-ghost-town-of-Pripyat-by-Michael-Day.html.

15 A. Vaughan, 'Wildlife Thriving Around Chernobyl Nuclear Plant Despite Radiation', *The Guardian* (5 October 2015). Available at: http://www.theguardian.com/environment/2015/oct/05/wildlife-thriving-around-chernobyl-nuclear-plant-despite-radiation.

Physics: 'How old are the peat bog people?'

As a way to introduce carbon dating, show photos of peat bog bodies which have been preserved for thousands of years and ask the students how old they think they are, before moving on to explaining how scientists are able to work it out by using radiocarbon dating. There are plenty of *CSI* type stories of bodies being found and radiocarbon dating being used to solve the mystery of their age. You could even make up your own murder mystery where students have to investigate how long teachers in the school have been dead using data from radioactive isotopes (choose the right members of staff here).

Whilst introducing a topic with a story is brilliant for the first lesson, obviously it will not be possible to teach the detail we need to get into in this manner. But the beginning of a lesson is always a good opportunity to link to prior learning and stories contextualise new knowledge in the students' existing understanding of the world. (There are a number of ways to review prior learning as a starter activity in Chapter 3.)

Chapter 2
IT'S ALL IN THE EXPLANATION!

'SO HEAT RISES ...'

'It's all in the explanation' sounds simple to the point of being simplistic, but in the busy life of a teacher it's easy to succumb to the temptation of thinking about and planning the tasks a class will complete without properly considering how you will actually explain or introduce a topic. I know of English teachers who script out or plan the questions they are going to pose to their class, but as science teachers we can tend to go into autopilot, assuming that because we understand a topic then that means we can adequately explain it to others without having to think about it too much.

To address this problem, you will need to know your students and what their initial level of understanding is likely to be, as explanations must be suitable for your audience. Then, with the end point in mind, plan out what steps the students will need to take to understand the topic – for example, if it's a tricky topic, review how a couple of textbooks attend to the issue. Explanations need to be clear, direct and easy to understand, and they need to be linked to opportunities to practise and check understanding before students move on to the next area of complexity. Visuals, props and multiple models can also help students to really get it (examples of how to weave this into your teaching can be found in Chapter 3). It's also essential that you make it part of your performance to constantly interweave the subject specific language into all conversations and that you plan how to act it out – considering how your tone of voice can be used to place emphasis where it is needed.

We'll also look at the importance of questioning in this chapter – good explanations are not a one-way street but a dialogue.

AVOIDING THE MISSING RUNG

If we fail to think about and plan properly for how students are going to learn, we can tend to present the task or topic in a manner that resembles a ladder with certain rungs missing (i.e. vital steps that students need to learn). Without all of the rungs, the students are presented with too much information at once which can lead to memory overload and, in turn, to disengagement or naughtiness. Breaking the learning down into small steps does not mean 'dumbing down' the curriculum; in fact, it serves to help them move forward with challenging tasks and gives you the opportunity to really think about how you are going to introduce a topic. It also helps teachers to anticipate and address the common misconceptions students might have.

When planning how students are going to learn:

- Start with the end point – what is it that you want them to be able to do at the end of the topic or lesson?

- Boil this down to the key concepts you want the students to grasp and the steps you will need to go through to achieve this.

- Think about the misconceptions the students may have and plan this into your steps (e.g. that respiration and breathing are interchangeable).

- Plan how you will assess the students at each rung and how you will provide them with feedback along the way so that you do not move up the ladder too quickly. Think about how you can engage the students in this process.

- Think of some memorable moments to make the science really come alive so the students don't get too bogged down in the detail.

- Don't forget to plan opportunities for practice at each step of the ladder so you can provide useful feedback as you go.

AN EXAMPLE: BALANCING EQUATIONS

Balancing equations can be a really tricky topic to teach – for some students it may well come to them completely intuitively, others may be able to learn how to replicate a method you have taught them, whilst the remainder won't have the first idea what you're going on about and will start chatting about football. The mistake here would be to teach balancing equations as an aside during a lesson on ionic bonding before it has been introduced properly. The suggested steps below are the proposed rungs of the ladder which could be taken over a number of lessons (e.g. with a younger group of students) or could be accelerated through more rapidly with students who are reviewing balancing equations, with the first four steps being part of the prior knowledge segment of the lesson. Avoid rushing at times like this as the misconceptions that could be picked up in the meantime will cause a lot of wasted time as soon as the examples become challenging.

Step 1: Start with simple word reactions and introduce these through simple experiments

Start with a simple experiment:

Hydrochloric acid + sodium hydroxide → sodium chloride + water

Or use a whizz-bang reaction like thermite to engage.

Next up: some guided practice for students, starting with scaffolded equations:

Hydrochloric acid + _____ hydroxide → calcium chloride + _____

Gradually remove more parts of the equations to scaffold the learning. Provide the students with an opportunity to work at their own pace by allowing those who have got the slightly easier questions to move on more quickly than others, provided they have demonstrated that they understand.

Step 2: Review previous learning – symbols and chemical formula

Check students are happy with chemical symbols (e.g. play bingo or chemical symbol splat – see Chapter 3). Ensure students understand what a chemical formula means, perhaps through a simple table:

CO_2	
Number of carbon atoms	1
Number of oxygen atoms	2

Or get students to make molymods of different compounds and link this to formulae. Also, review the idea of the molecule because a common misun-' derstanding students have is why H_2 is different from a hydrogen atom. If this is not addressed at this point, there will be confusion later on.

Step 3: Introduce simple symbol equations that do not require balancing

This could link back to Step 1 if appropriate, or get the students to write symbol equations without balancing. Provide opportunities for practice here and bring in molecules (as mentioned above) as there will be simple misunderstandings such as when to use capital or lower case letters.

Step 4: Conservation of mass

Explain the principle of the conservation of mass using props rather than a PowerPoint slide – for example, burn iron wool to show that its mass increases when it becomes iron oxide or get the students to complete a practical burning magnesium in air. Model the process using Skittles, Lego or the classic molymods to demonstrate that reactants all become products.

Step 5: Simple balancing equations

Only at this point are you ready to teach balancing. Start with simple equations that only require using the number 2 before the symbol.

The common misconception here is that you can change the formula of a compound rather than the ratio of the compounds/elements. Ensure that you teach this explicitly to the students (again using diagrams or molymods to illustrate). One way around this is to get the students to draw the molecules before they start and use that as a method for working out the balance. This will not help them to answer hard questions but it will support them in understanding the abstract idea. In doing this (as in the diagram on page 25), the students will appreciate that there would not be enough oxygen on the reactants side to make the products on the right; therefore the only thing that could be done would be to increase the number of reactants by doubling the number of oxygen molecules. Another more imaginative way here would be to go to the PE department and ask to borrow some coloured bibs, so the students can act as different atoms and you can use them to model the molecules instead.

$$CH_4 + O_2 \longrightarrow CO_2 + H_2O$$

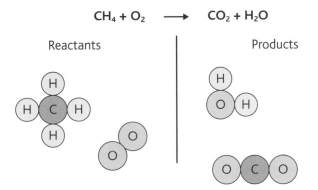

Reactants | Products

Step 6: Challenging equations

Now that the students understand the concept behind balancing equations, you can teach them a tool to help them balance more difficult questions.

The table method

Start by identifying the number of atoms in each compound/molecule. In this example, the students need to be told to start on the reactants as they have the smallest number of atoms and you can only make numbers bigger to balance. What can they do to the coefficients to make iron balance first? Students would most likely start by adding a 2 in front of the iron, which of course would not balance the equation, so they will need to think again. Continue to work through possibilities using the table until the equation is balanced.

$$4\,Fe + 3\,O_2 \rightarrow 2\,Fe_2O_3$$

Element	Number in Reactants	Number in Products
Fe	~~1~~ 4	~~2~~ 4
O	~~2~~ 6	~~3~~ 6

Now the students need to practice, so give them plenty of opportunities to repeat the steps above.

Step 7: Never forget

Now that the students understand the principle behind balancing, don't let them forget. Get them to balance equations at every occasion, such as when teaching ionic and covalent bonding, during rates of reaction and so on. Each of the steps should contain opportunities for students to practise and master each step *before* they move on – guided practice is essential if they are to master balancing equations.

This is just one example when it can be really easy not to plan exactly how you are going to teach (there are many others), but it really pays off to take a step back in your planning before you start thinking about the exact tasks or starters that students may complete in a lesson.

HOW DO YOU KNOW YOUR EXPLANATION HAS BEEN GOOD ENOUGH? THE IMPORTANCE OF QUESTIONING

Questioning is a skill: asking the right questions to the right students can help them to make giant leaps in understanding. Whilst a class Q&A may be the first method we tend to reach for, it is not always the best – those students who put up their hands will already know the answer and those who don't, don't bother. Thinking about the key questions you would like the students to answer during your planning and then employing methods where all students are engaged, rather than just being passive consumers, can have a big impact on understanding.

Think–pair–share

Pose a question to the class, give them 60 seconds of thinking time in silence, followed by a further 60 seconds to discuss their thoughts with a partner, before asking them to give feedback to the rest of the class (perhaps try using a student selector tool to keep them all on their toes). You may find that providing students with a word frame or written questions may support their discussions.

Here are three ideas for using think–pair–share:

1   Questioning after a demonstration. Give the students key questions to think–pair–share, providing sentence starters if you deem them necessary. Keep the questions as open as possible. This can also work just as well after the students have completed a practical that has provided anomalous results.

2   The emotive starter. The link between emotions and learning is clear: if you have an emotive response to something you are more likely to remember it.[1] An emotive starter is a good way to harness this. Display a photo of something distressing – this could be a photo of someone from Hiroshima suffering from radiation sickness or a global warming related drought or flooding event. Then pose the five W's: Who? What? Why? When? What next?

3   Concept cartoons. There is a wealth of these on the Internet – simply search for 'concept cartoon' and your topic.[2] They are particularly useful when there are a number of misconceptions around the topic in question. Using the think–pair–share process, the students should choose the student (or statement) they most agree with and justify their choice.

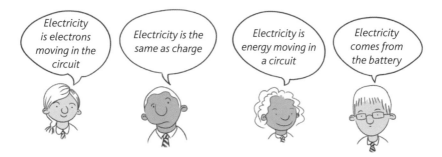

---

1   See W. Harlen, *ASE Guide to Primary Science Education* (Hatfield: Association for Science Education, 2011) and L. Pessoa, 'On the Relationship Between Emotion and Cognition', *Nature Reviews Neuroscience* 9 (2008): 148–158.
2   A good place to start is: https://www.stem.org.uk/elibrary/resource/26497/concept-cartoons.

Snowball

Start this activity in the same way as a think–pair–share but 'snowball' this into groups of six, within which the students have to agree on an answer. Add a further level of complexity by getting the snowball group to come up with one significant question to ask back to the class or teacher.

Mini whiteboards

Mini whiteboards are good for short closed questioning when you have a number of key words or ideas that you want to put across or check for understanding.

Continuum line

Use a continuum line if you want to gain the views of a number of students independently of one another. Stretch a ball of string across the room, with one side representing 'agree' and the other 'disagree'. Ask questions about your topic and then get the students to move to the area of the string that best illustrates their viewpoint. Choose a few students at random points along the line to feed back to the class. Then, once the students have listened to one another, ask if any of them want to move (i.e. to alter their position on the question posed).

Another way of using the continuum line is to ask them the question, 'Hearts or heads?' Once they have made their decision, they either put a hand on their heart if they have had an emotional response to the question or a hand on their head to indicate that they are using facts and understanding in their decision.

Ball games

There's a lot of learning to be had from having a blown-up beach volley ball readily to hand – but beware of health and safety issues.

Get the lot of them on their feet. Ask a question of the class and throw the ball to the first student who indicates that they know the answer. They then ask another question and pass the ball to the next student who can answer. Those who have answered a question can sit down. Continue until the whole class are seated.

Alternatively, split the class into two. This time the two teams play each other, asking questions to a member of the opposing team. If the other student can't answer the question then they have to sit down. The winning team is the one left with students still standing.

Both of these techniques allow you to do the one really important thing – listen.[3] The answers the students give are like the results section of the science reports you did at uni – they mean nothing unless you have thought about them and planned what to do next. Have the students got it? What might you need to go through again? The quicker you can respond to this, the better. If you can deal with it then and there, then great – do it. If not, revisit it during the next lesson.

---

3   There's a life lesson here.

Chapter 3
MAKING IT MEMORABLE

'I THINK WE DID THIS LAST YEAR ...'

The teacher I remember the most from my own schooling was my Year 8 chemistry teacher, Dr C, who delivered memorable moments daily. Unlike the rest of my science teachers, who preferred to use the overhead projector and workbooks, Dr C spent most of the lesson getting various students up to the front of the class to act as atoms. I remember so many of his lessons for the simple reason that they were memorable, and I attempt to replicate this principle in my own teaching.

The best way of getting brilliant ideas on how to teach things is through discussion with your colleagues, reading blogs and consulting websites. In my school, a 10 minute morning meeting is timetabled once a week so that we actually get to share ideas with each other. This stops our conversations revolving around behaviour rather than learning.

Making learning memorable is important because most people do not remember most of what they learn. They may do in the short term, and will be able to show you that learning in a plenary. But will they have retained the knowledge in a week's time? A year's time? Even when they have learnt something, sometimes they can't apply the knowledge to a new situation or use it usefully in the future. Partly this is because the brain very quickly reaches cognitive load and decides what it is important to remember and what it isn't.[1] (The classic example of this is the video by Daniel Simons and

---

1   See J. Sweller, 'Cognitive Load During Problem Solving: Effects on Learning', *Cognitive Science* 12(2) (1988): 257–285.

Christopher Chabris where viewers are asked to count the number of times the basketball players pass the ball, whilst at the same time missing the man dancing in a gorilla suit.[2]) Also, we often assume that because students have learnt something means that they can 'transfer' their knowledge from one area to another when, in reality, we have the responsibility to ensure that we provide opportunities for students to practise this skill.[3] We must provide vivid learning opportunities, routed in prior learning, with opportunities to teach how to think through a new problem in order for students to learn to transfer knowledge.[4]

Tapping into students' emotional landscapes can help teachers to make something memorable. Stories help here, as does the way in which you order the material. But how about the way in which you present the information? If your lessons always follow the same structure or always involve the same tasks, then they will quickly become boring and, by implication, unmemorable. Fun, surprise and change can glue something in a student's mind so they still remember it in five years' time.

In this chapter I have gathered together some diverse ideas that you can use to provide rich experiences for students. By this, I do not mean tasks that operate solely for the pursuit of fun (although, of course, some do) but those that really move the students forward if used correctly. Most of these methods work in the same way: by making an idea, which might otherwise be viewed initially as being a little dull or uninspiring, become exciting – and

---

2 The video can be found at: https://www.youtube.com/watch?v=vJG698U2Mvo.

3 See V. Strauss, 'The Real Stuff of Schooling: How to Teach Students to Apply Knowledge', *Washington Post* (24 March 2015). Available at: https://www.washingtonpost.com/news/answer-sheet/wp/2015/03/24/the-real-stuff-of-schooling-how-to-teach-students-to-apply-knowledge/.

4 The Education Endowment Foundation has found that teaching metacognition strategies is one of the most cost effective and high impact strategies for improving student outcomes. See: https://educationendowmentfoundation.org.uk/evidence/teaching-learning-toolkit/meta-cognition-and-self-regulation/.

thereby enabling students to convert them into memories which are colourful, visual and easy to remember. However, beware of using these ideas as 'fillers' that just keep the students busy. Always start with the premise: what are they going to learn from this?

In my NQT year, a colleague shared a rule with me that I still employ, which is to do one 'Wow' lesson with a class each week where you really go to town on planning something great. This means that if you then spend one lesson working on writing skills, they are less likely to view science as dull. This rule makes lesson planning manageable as no NQT is a better teacher for working themselves ragged.

SING WHEN YOU'RE WINNING

Silly songs can be a really useful way of getting pieces of information into students' heads in a memorable way. In terms of science teaching, these are best used for those topics where students get confused between two things (e.g. alkanes and alkenes, mitosis and meiosis).

Songs can be made up quite quickly and you can find loads online.[5] I like to marry the words to a well-known tune – nursery rhymes or recent pop hits work equally well. Use a few extroverted students to help you demonstrate this to the rest of the class the first time you introduce it and use background music so they aren't too embarrassed to join in. After one round insist that all the students get involved – and they will! Make sure you use the songs over a number of lessons and the learning will stick.

---

5   YouTube is your best bet. Check out these channels: ParrMr, Mr Lee and acapellascience2 (especially 'All About That Base (No Acid)'). OscTV's 'The Periodic Table of Rock' always goes down well, as does Emerson Foo and Wong Yann's 'Electromagnetic Spectrum Song' (there's a karaoke version of this one!).

Here's one I made up earlier to the tune of 'Happy' by Pharrell Williams:

It might seem crazy what I'm about to say:
Mitosis is here, so you can grow away.
It'll help you regenerate, repair and grow.
Without it you'll struggle so keep going by the way
Uh.

Because I'm MITOSIS:
Clap along if you feel like cells which don't change.
Because I'm MITOSIS:
Clap along if you feel the daughter cells are all the same.
Because I'm MITOSIS:
Clap along if you know what diploid is to you.
Because I'm MITOSIS:
Clap along if you feel it's happening inside of you.

Here comes the next one, don't get too confused –
Well, meiosis is here to make you your egg and sperm.
Well, it'll halve your DNA and chromosomes, yeah.
No offence to you, but you need some difference.
Here's why.

Because I'm MEIOSIS:
Clap along if you feel like you're cells that like to change.
Because I'm MEIOSIS:
Clap along if you feel that the daughter cells are *not* the same.
Because I'm MEIOSIS:

Clap along if you know what haploid is to you.

Because I'm MEIOSIS:

Clap along if you feel it's happening inside of you.

This works in the same manner as any mnemonic we might use to help us learn: by providing a cue (in this case a well-known tune) to help the students to remember. Even just using music related to the topic as the students walk into the classroom can be a good way of getting them excited. I use 'Chemistry' by Girls Aloud, 'High Voltage' by AC/DC, 'I've Got the Power' by Snap, 'Right Here, Right Now' by Fat Boy Slim (mainly for the evolution video) and 'Radioactivity' by Kraftwerk.[6]

A final idea here is to get students to come up with their own. And, yes, you can write it as a rap!

DANCE IF YOU WANT TO LEARN FASTER

Much like using song as a memorable way to learn, dance can be used too. Dances work best when you are teaching topics that require the students to learn a chain of events. Topics such as how hormones function in the menstrual cycle or how coal is formed work brilliantly.

---

6   My editor, who obviously has great taste in music, also recommends 'Electricity' by Captain Beefheart and 'Inertia Creeps' by Massive Attack.

This dance is good for learning how fossil fuel power stations work.

Wiggle fingers
(crouched down)
= burning
fossil fuel

Move upwards
wiggling fingers
= steam rising

Roll arms
= steam moving
turbine

Arms move in
and out
= generator

Jump up
= step up
transformer

Glide along
= national grid

Jump down
= step down
transformer

Pour a cup of tea!

I also get students to record their dances and then use this later during revision (or in later years just to embarrass them – emotions are important in learning, after all!).

SILENT DEBATE

An inspirational English teacher taught this one to me, and it is perfect for some of those trickier topics in science when you really want students to listen to different points of view. On a sheet of A3 paper, write down emotive statements about the topic (e.g. 'Alcoholics should not be given organ transplants', 'People should be automatically put on the organ transplant list') and place them around the room. Make sure that the students maintain complete silence and that they have different coloured pens so you can differentiate who wrote what. In groups, the students move to a statement and write down their opinion about it. They then move clockwise to the next statement. At the next statement, get them either to write a new opinion or respond to the opinion already written by the previous group. Continue until each group has been to each statement. Choose a student from each group to feed back some of the conflicting arguments to the class.

ROLE PLAY

For some practical technical tips on how to best do a role play activity, go and watch the best drama teacher at your school – they are masters at this. Role play can be used in two distinct ways in science.

First, more traditional role playing techniques can be used when debating scientific topics in which students are given a role in the debate. Topics this would suit range from stem cells, genetically modified food, nuclear power, the use of animal research and nanotechnology to whether we should all be vegetarian. Good ways to do this include:

- Hot seats.[7] Get a student to represent a particular viewpoint (e.g. a farmer opposed to genetically modified crops) and take a seat in your chair at the front of the classroom. The rest of the class then ask them questions on their opinions which they have to answer in character (depending on the class you may need to provide them with some background material to support their characterisation or give them some research/planning time prior to taking up the hot seat). Continue with other students who represent different stakeholders (in this example, perhaps the CEO of a food company, a child from a developing country, a pensioner living on a tight budget or an environmentalist).[8]

- Debating ring. Set up your room with the students sitting in a square (move out of your lab if possible). Put the students into two opposing teams (blue and red, if you will), one for and one against the house's statement. Give the teams time to prepare as many arguments for their view as they can think of, so that ideally each student in the team has their own unique argument, perhaps written down. They are then ready for the debate! Get one student at a time from each team into the blue or red corner. In round one, a student from each team comes to the ring and states their point. The teacher (or, better still, a student

---

7   You can also use the hot seat as a good plenary. A student sits in a chair at the front and is asked questions based on the current topic (either by you or the class). If they hesitate or get the answer wrong then they are replaced by another member of the class. Announce a winner at the end. The hot seat works best if you have a nice big leather chair – you can then play the *Mastermind* theme as the student takes to the seat which adds a nice bit of drama!

8   For more ideas read M. Francis, 'The Impact of Drama on Pupils' Learning in Science', *School Science Review* 327 (December 2007): 91–102. Available (for members of the Association for Science Education) at: https://www.ase.org.uk/journals/school-science-review/2007/12/327/.

referee) decides which team has won based on that statement and they are awarded a point (or a 'knock-out' if the point is that eloquent and just). Next, another pair of students come to the centre and the process is repeated. This continues until one of the teams runs out of arguments and the wining team is announced.

Of course, you could just use a more traditional debate set-up but you're unlikely to be able to involve as many students in it and it won't be quite so much fun. The BioEthics Education Project (BEEP) is a good resource for case studies and videos to support discussion.[9]

A word of caution: with so many sensitive topics to teach, make sure you know your students as it can be easy to miss the student who was conceived by IVF or the student whose parent is an alcoholic. If you have a new class, identify the topic at the end of the previous lesson and tell the students that if they are uncomfortable with the upcoming content to speak with you before the next lesson.

The second way role play can be used is to model scientific phenomena. In essence, this involves using your students to act as electrons, DNA bases, gas molecules and the like. Here are some ideas:

- Modelling protein synthesis (because you'll probably find that you have to teach this about five times before the students really get it). Get some of the students to be DNA bases, some mRNA and so on.

- Acting out the different states of matter. Split the class into three, ask each group to model one of the three states and then get them to show their role play to the rest of the group.

- Displacement reactions. Give one student a crown and they act as the prize (e.g. a potassium prize) that the most reactive halogen gets to win

---

9   See www.beep.ac.uk.

or react with. Ask a few others to act as halogens. Get the rest of the class to predict if a reaction will occur. Write up the equations.

- Rates of reaction. Get the students to be particles and ask them to walk around the room, occasionally bumping into one another to model a collision. To demonstrate the effect of a higher temperature, tell them to move faster and knock into each other more frequently (make sure you remind them of health and safety first). For pressure, stop them from moving around the whole of the classroom to just half of the room. For concentration, compare the collisions if half of them are sitting down and not involved to when they are all involved. Finally, to demonstrate the use of a catalyst, set up some fabric or string across half of the classroom which all of the particles are 'attracted' to.

USING PROPS

Have some props on the front desk at the beginning of the lesson and ask the students to work out what the lesson is about – for example, you could have some leaves, some caster sugar, a light bulb and a bottle of water.[10] Ask the students to link the props in order to guess the lesson topic. This is great to use as a reminder of previous learning when revising or revisiting.

---

10 The answer would be photosynthesis.

SCHRÖDINGER'S CAT

This is a game for taking props a step further. Start with your items in a cardboard box at the front of the room. The students can ask questions until they work out what is inside the box (you may need to give them some hints). You can also teach them about the concept of Schrödinger's cat whilst you are at it – that a cat contained in a box alongside a radioactive source (which, of course, will decay randomly) is considered by quantum mechanics to be both alive and dead simultaneously until it is taken out of the box and observed.[11]

ANALOGIES

We understand new things based on what we already know so abstract ideas can be difficult to grasp. An analogy can really help the students to make meaning of a new topic. Analogies need to be used alongside other robust methods as they can sometimes lead to misconceptions. Here are some good examples:

- Water pipes as an analogy for how electricity moves through a circuit and how resistance affects current.

- Comparing the structure of the atom to the structure of the solar system.

- Evolution as a tree.

---

11 If physics isn't your strong point then consider reading John Gribbin's *In Search of Schrödinger's Cat* (London: Black Swan, 2012).

- Analogies to explain the size of an atom – if a hydrogen atom was the same size as a golf ball, then a golf ball would be as large as the Earth.

A comprehensive list of examples can be found at www.metamia.com.

THE WORDLESS ESSAY

The wordless essay is a great way for students to illustrate their under-standing without necessarily having to write a full A4 page about it, and it works really well as an alternative to the students writing up a lab report. For example, if you are completing a practical into the effects of acid rain on the germination of cress seeds, the students could use cameras (camera phones are great if your school allows it) to take photos of how they set up the experiment. They then take a picture of the growth of the cress seeds each day for a week or so. The challenge for the students is to ensure that everything they would have written in an essay is portrayed in the photos (e.g. how much acid they measured out, how many repeats they did, how they controlled the sunlight). The students will probably come up with some imaginative ideas here, but captions written in speech bubbles are both common and useful.

Once the practical is complete, the students should compile their photos using software (e.g. Windows Movie Maker or PowerPoint) and present them to the class for peer feedback or hand in to you for expert feedback. I some-times refer to this as the 'Instagram essay': for some reason the students think it is more fun that way, particularly if they can use filters. They can even make a Flipagram of their photos if you're allowing them to use their phones.

Similar results could be achieved by the students producing a cartoon reel (the website PowToon is great for this – www.powtoon.com). Another (and very 21st century) method would be to get students to explain what is going on in a process using only emojis (google 'Bill Nye and emoji' for an amazing example of how to explain evolution through emoticons).

CHOICE BOARDS

A choice board does what it says on the tin: rather than giving all the students the same task in order to fulfil the same objective, a choice board gives them alternatives to choose from. Make sure you know your students' strengths and weaknesses as you might have to nudge them in the right direction. Using a similar method with homework (often referred to as take-away homework) can seriously decrease the number of detentions you have to hand out.

GROUP WORK

One thing that students comment on when they transition between primary and secondary school is that the amount of group work and discovery learning drops dramatically. Many teachers are wary of using group work, partly because it can be difficult to truly appreciate whether learning outcomes have been reached and, thus, if the lesson has been successful or not. However, research performed by the SPRinG (Social Pedagogic Research into Group-work) project has found that students who engaged in grouped activities throughout Key Stage 3 made greater progress in subsequent science exams than the control group students.[12] A further study by the Wellcome

---

12   See www.spring-project.org.uk.

Trust into students' attitudes towards science also demonstrated how vital students felt the opportunity to interact and participate was to their learning and engagement.[13]

Employing group work regularly can develop communication skills and allows students to explore alternative answers that, in turn, support them as they attempt to solve cognitive based problems. However, there's no doubt that group work can be a slow burner and it can be difficult to see 'progress' (should such a thing actually be observable) in a lesson when group work is used. It will develop skills over time but make sure you embed key words and ideas either before or after as, at times, it can result in rather vague explanations of phenomena that lack the level of detail required for students to be successful in science.

A number of the ideas outlined here require group work, which can be a challenge with classes at the beginning of the year, but when well thought out and used regularly group work can result in great gains for students.

Here are some ideas to think about before you start.

- Always plan your groupings in advance and think about the balance of students in each group.

- Consider if ability or mixed ability groups will be best. Ability grouping allows you to easily differentiate resources, whereas mixed ability allows different students to bring alternative strengths to the activity.

- How will you organise the classroom? Where will the groups be located to minimise noise? How will you ask them to move around the room?

---

13   National Foundation for Educational Research (NFER), *Exploring Young People's Views on Science Education: Report to the Wellcome Trust* (2011). Available at: http://www.wellcome. ac.uk/About-us/Publications/Reports/Education/WTVM052735.htm.

- What roles will you give to each student? Possibilities include: chairperson, time-keeper, scribe and reporter. Inform the students who are the chairperson for each group that their role is to ensure that every person in the group is given an opportunity and time to discuss their point of view or idea.

- How will you set out ground rules for discussion?

- What will you do to ensure that all students participate? One idea is to give a number to each student and then, during feedback, call on all number ones to feed back on one area, then all number twos on another and so on to keep them on their feet. If you are getting the students to record their discussions on sugar paper, make sure they are using different coloured pens so you can see from a distance that they have all contributed.

- How will you know the students have moved forward in terms of their skill development and understanding of the task? How will you check that they haven't made errors in their thinking?

Here are some suggestions to facilitate group work.

Ask the expert

Ask the expert is a lovely way for students to collaborate when they're revising. Split the class into groups of five or six and give each group a different topic to research and revise, asking them to produce 10 key ideas that all the students need to know. After about 15 minutes, choose an expert from each group. The expert remains at their table whilst the remainder of the group separate and visit one other expert each around the room (when they visit an expert there should be only one representative from each group present).

The expert should then go through their topic with the group (you will need to give them some instructions on what makes a good expert to avoid them just reading out their notes). The students then return to their original group and have one or two minutes each to go through what they have learnt with the rest of their group.

Researchers

In groups of five or six, give each student a different piece of information about a topic – this could be a newspaper article, a section of a textbook, a printout from a science blog or research paper (for more able students). Give the students five minutes to read the article, make notes and record any questions. Then give them two minutes each to present their findings to their group. To complete the group task, pose some questions:

- Which of the pieces of evidence do you think is the most reliable? Why?

- Do you think there is any bias in any of the articles?

- What more do you need to know before you can make a conclusion?

GUIDED STUDENT PRACTICE: METACOGNITIVE APPROACHES

If you want students to think like scientists, to analyse data and come up with their own conclusions, then it can be a useful trick to speak out loud your own thought processes when demonstrating how to attempt a difficult question. This generally goes by the name of modelling but it is entirely reasonable to put a metacognitive tag on it. For example, metacognition can be used to help students understand how to explain the complex results of a practical. To model this, a teacher would explain their thought processes whilst they are looking at the data and how their thinking results in or leads to a conclusion. This will need to be done repetitively over a number of lessons to ensure understanding because teaching higher level skills takes substantially longer than simply pushing key words into their heads. Teachers who take time over modelling problems (and the solutions to them) provide students with the tools they need to work independently; teachers who skirt over brief explanations can result in stuck students.[14]

MEMORY DEVELOPMENT – BACK TO THE BIGGER PICTURE!

With all of the subject knowledge that has to be taught, it is super-easy for students just to forget every single thing you've taught them if you're not vigilant, even if you think you've made the lessons memorable. There is no point berating them for this, but it is well worth thinking about whether your teaching style has caused (or accepted) this forgetting.

If you've ever tried learning a new language from a CD or audio guide, you'll probably already be fully aware of spaced repetition: the planned, spaced out reminders of previous learning. Any student who has spent too many late nights trying to get to the next level of *Call of Duty* will also understand the importance of repetition themselves. This is not in any way a new concept,

---

14 See J. Dunlosky, K. A. Rawson, E. J. Marsh, M. J. Nathan and D. T. Willingham, 'Improving Students' Learning with Effective Learning Techniques: Promising Directions from Cognitive and Educational Psychology', *Psychological Science in the Public Interest* 14(1) (2013): 4–58.

but it is one that can be easily forgotten about in the middle of darkest November (the hardest month for any NQTs or PGCE students – it's dark, it's not yet Christmas and you're still building relationships). It's likely that you already employ spaced learning to some degree or another but making it explicit in your teaching really helps with knowledge retention.

THE RETRIEVAL PRACTICE EFFECT

In 1885 Hermann Ebbinghaus, a German psychologist, came up with the idea of the 'forgetting curve' to describe how memories could possibly be prolonged. The main point of the forgetting curve is that we remember the ideas that we have learnt recently or that we've recalled regularly. This is sometimes referred to as retrieval practice. To ensure our students' recall of information is at its optimum, we must plan in opportunities for them to recall every distinct piece of knowledge taught on a variety of occasions to stop them forgetting it before they can use it again.

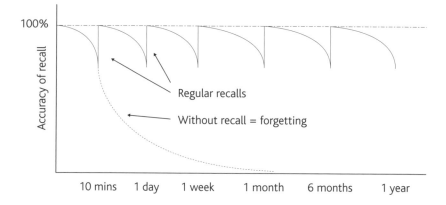

PRACTICAL WAYS TO REVIEW AND REVIEW AND ...

Retrieval practice goes far beyond the plenary and the mini-plenary. Use this whenever you are aware that there is a possibility of factual overload in lessons. There are certain topics that retrieval practice is particularly well suited to, including much of plant biology, waves, patterns in the periodic table to name but a few. Research undertaken into the use of regular, spaced quizzing of students proves that not only does regular quizzing result in large memory gains, but it also promotes the flexible use of content that could later be useful in an exam.[15]

As classroom practitioners, and as leaders of learning (which is a distinct skill from being a researcher), variety can be an important way of keeping students engaged – not only do the students not want to do the same task every day, but neither do teachers want to continually teach using the same lesson format. This is not the same as making the students do loads of massed practice, as can be common in some Asian countries. What we try to do is vary the ways in which we present the knowledge as this will result in the students being better able to apply the knowledge they have (one of those essential skills frequently discussed but so difficult to develop in students in a short period of time).

---

15 See M. A. McDaniel, P. K. Agarwal, B. J. Huelser, K. B. McDermott and H. L. Roediger, 'Test-Enhanced Learning in a Middle School Science Classroom: The Effects of Quiz Frequency and Placement', *Journal of Educational Psychology* 103 (2011): 399–414; and H. L. Roediger and J. D. Karpicke, 'Test-Enhanced Learning: Taking Memory Tests Improves Long-Term Retention', *Psychological Science* 17(3) (2006): 249–255.

USING THE FORGETTING CURVE IN THE CLASSROOM

I have described below some of the ways in which you can harness the idea of retrieval practice (or quizzing) in the classroom with a little more creativity. I have focused here on the parts of a leaf – a topic which can quite easily be taught just once and then left well alone. There are numerous new key words for the students to learn and many of these are not linked in any obvious way to other areas of the curriculum. This makes it imperative that retrieval practice is enforced as they will not come up against this learning again in any other curriculum area. These strategies work nicely after a topic has been taught but they should not act as a substitute for teaching the topic in depth to begin with. There is little point in using these just for memorisation, as learning isolated facts will not allow a full understanding of any concept.

Most of these techniques involve the learning and understanding of key words to enable further thinking and many also provide instant feedback to students. Most also revolve around team work and competition, which can be great ways of getting those students who dislike 'practice' involved in the necessary repetition of content without even noticing they are doing so. After all, the reason that students continue with the same computer game, or adults try to learn another language, is based on motivation to reach a learning destination. Just occasionally (!) in education we have to help students along that path, so competition, variety and fun can really help. On a final note, don't worry too much about the amount of time you find yourself spending on this – in the long term the gains produced by easily recalled learning are well worth it.

These techniques provide six opportunities to touch on the topic over the space of a single half-term. Of course, by definition, this involves going over the same content again and again, but with the right tools this need not be boring.

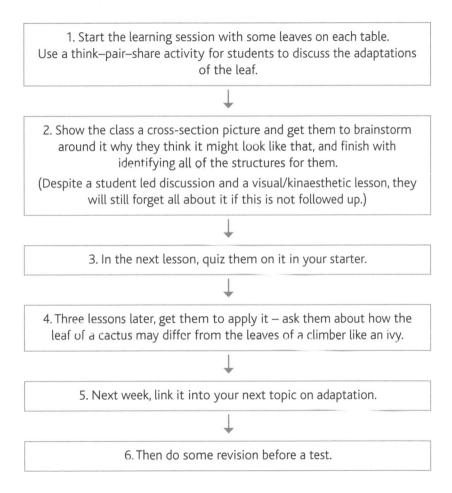

1. Start the learning session with some leaves on each table.
Use a think–pair–share activity for students to discuss the adaptations of the leaf.

↓

2. Show the class a cross-section picture and get them to brainstorm around it why they think it might look like that, and finish with identifying all of the structures for them.

(Despite a student led discussion and a visual/kinaesthetic lesson, they will still forget all about it if this is not followed up.)

↓

3. In the next lesson, quiz them on it in your starter.

↓

4. Three lessons later, get them to apply it – ask them about how the leaf of a cactus may differ from the leaves of a climber like an ivy.

↓

5. Next week, link it into your next topic on adaptation.

↓

6. Then do some revision before a test.

Here are some more techniques that are quick, resource light and perfect for recapping topics.

Splat

This comes first because it is a favourite of every single class I have ever taught! You need one PowerPoint slide with around 15 key words linking to your topic projected onto your whiteboard (if you write them on a board they will get messed up during the game). For example:

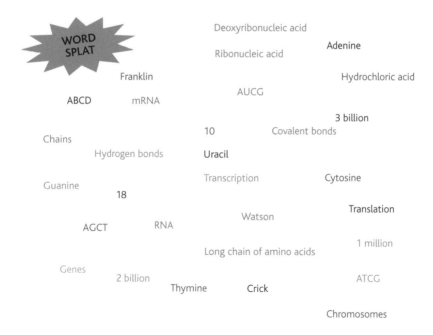

Split the class into two groups. Name the groups using two opposing words from the topic you have been teaching (e.g. Gamma vs. Alpha, Mitosis vs. Meiosis or Electron vs. Proton). You will need one representative from each group at the front, one stood on either side of the board.

Read out a definition of one of the key words displayed on the board. The first student to find the key word and then hit (or splat) the word with their hand, at the same time as saying 'Splat!', wins the point for their team. The student who misses is replaced by another member of their group. I tend to play first team to five wins. Once your class have got the hang of it, provide bonus points to students who can come up with the definitions. Mix up the key words over various topics to review learning over a stretch of time.

Only connect

Only connect is great, particularly for revision. Divide your class into teams of between four and six students and get each team to put a list of words (like the one below) into families of four and justify their reason for doing so (e.g. they all involve ultrasound). The winning team is the first to get all of the words into families with justifications. Give one point for each group of four and a bonus point for each category. (You can also use the same resource later in the year to play bingo.)

Infrared

Hippos

Bats

Earthquakes

S-waves

Ultraviolet

Volcanoes

Toothed whales

Elephants

P-waves

Dolphins

Radio

Moths

Microwaves

Seismic waves

Epicentre[16]

Articulate

Articulate is a favourite game in homes across the country on Christmas Day and it is just as fun in the classroom – it is very similar to the game *Taboo*. Write down key words onto flashcard sized pieces of paper (you'll need about 20 different key words) which you then split into two piles. Divide the class into two (if you use the same groups for a half-term you can really build up some rivalry). You will need one spokesperson per team.

---

16   It's nice here to include words that could fit into more than one category to promote some discussion. Example answers: (1) Infrared, ultraviolet, microwaves, radio waves – all part of the electromagnetic spectrum. (2) Bats, dolphins, moths, toothed whales – all emit ultrasound. (3) Elephants, hippos, earthquakes, volcanoes – all emit infrasound. (4) Epicentre, seismic waves, p-waves and s-waves – all involved in earthquakes.

Give each student two minutes to articulate the words on the cards without using said word (ideally use an electronic stopwatch on the board to add to the tension). The student's team have to work out from the description what word is on the card – they get one point for each correct answer. The winning team is the one that gets the most points in the two minutes.

After some practice, the students should be able to make these cards themselves and play the game in their tables of six. Provide them with stopwatches.

Dough!

Students need Play-Doh or Plasticine for this activity. Put the students into groups of three and give each group a different set of key words. One person in each team is appointed leader – their role is to get their team to guess as many of the key words as possible as these are represented in Play-Doh by the leader within 60 seconds. The next team then does the same. The game continues with another member of the team becoming the leader and the game is repeated until all members of the team have been the leader and/ or all of the key word cards have been used. The winning team is the group who guesses the most in total.

Odd one out

You can either use four pictures on a PowerPoint slide or you can choose to read out four key words (e.g. kinetic, potential, voltage, electrical; stem cells, cloning, mitosis, genetic engineering). In teams or pairs, get the students to identify the odd one out and explain their decision – crucially, they have to provide a reason for each one of the four words to be the odd one out. Use mini whiteboards to get feedback from all the teams at once and award students one point for each correct family of words and one point for explaining why they are linked. Keep track of the points given to the students to track how well each team is doing.

Name me

Produce a list of questions which all begin with the phrase 'Name me' (about eight works well). Put the questions upside down in the centre of the table. After a count of three, give groups of students a set amount of time to come up with as many answers as they can. Announce a winning table at the end.

20 questions

Sit a student at the front of the class in the 'hot seat'. Write a key word on the board behind them. They now have 20 questions to guess the word by asking their classmates questions about it. Alternatively, to engage more students actively in the activity, put them into groups of around six students. Get each student to think of a word related to the topic being studied, write it on a piece of paper, fold it up and put it in the centre of the table. Once each member of the group has done this, each student should select a word and, without looking at it, stick it to their forehead (with sticky tape doubled back on itself). They then have one question each at a time to ask the others about their word. Keep going around each student until all the words have been guessed. This is a great activity when you hit revision time.

Beat the quiz master

Put students into pairs (students of similar ability work best). Give them seven minutes to individually devise a quiz on the topic being studied – this could be a subject taught a week previously to enable them to benefit from some spaced learning. Get the students you have paired up to swap quizzes; then, once done, the tests are returned and marked by the quiz master. With classes preparing for exams, this can be developed further so they have to produce a past paper-style exam question.

There's an equation for that!

This is a useful plenary as it gets students to summarise the lesson in as few words as possible. Here are some examples:

Brain + adrenal glands + ovaries + testes + pancreas = endocrine system

FSH + LH + oestrogen + progesterone = female sex hormones

Can you catch out the teacher?

Misconceptions can be a real issue in science, so where you know that they exist, this can be a great activity. You will need to prepare in advance a bad answer like the one below:

> Respiration is breathing. You breathe in oxygen and breathe out carbon dioxide: these gases travel around in your blood capillaries to the muscles. Anaerobic respiration happens when you do lots of exercise and build up lactic acid. You then breathe out the lactic acid, which is why you breathe heavily even after you have finished exercising.

Get the students to list all of the mistakes – see who can get them all.[17]

Whiteboard challenge

This is a great plenary (it sometimes overexcites students so I tend to avoid using it at other times). Each student will need their own mini whiteboard and they should be grouped into teams of around six. The team must remain silent at all times!

Read out a definition for a key word. Individually, the students have to write down what they think is the key word being described without letting the others in their team see what they have written. When everyone has noted down their word, signal that they may now share the words with their team.

---

17 Respiration is *not* breathing (this is my utter pet hate!). It is a chemical reaction within the cells of the body. Anaerobic respiration does not necessarily result from lots of exercise (this is why when I go for long walks I still can't get rid of my muffin top and why gym buffs love high impact training). Finally, you don't breathe out lactic acid; you piss it out (well, you pee out the water produced from breaking it down).

The teams in which everyone has written down the same key word on their boards win a point. Keep a tally and, at the end, the team with the most points wins. (Without the competitive element it can work as a nice reviser at the beginning of the lesson; it's just a little less fun that way.)

The checklist

Checklists can be used productively to identify strengths and weaknesses and to support you in planning your future lessons. But it can also be a complete waste of time if some students don't take them seriously and write down 'amber' for every item.

Here is an example of a typical checklist:

Topic: Energy	Red/Amber/Green
I know that work done is energy transferred	
I understand the difference between work done and power	
I can calculate energy and power and rearrange the equations	
I can calculate kinetic energy	
I can calculate potential energy	
I know that energy is always conserved	
I can describe kinetic and potential energy in a real life example	

Here is a better example, in which students have to prove that they can do each step:

I can ...	If can ... then I can do ...	Prove it (students complete this section to show they can actually do this)	Red/ Amber/ Green
Describe the difference between work done and power	Give one example of work done and one example of power	Work done: the energy transferred (e.g. to move my chair 5 metres) Power: the work done needed to move my chair 5 metres, taking into account how long it took me to do it	Green
I can calculate energy and power in complex problems and rearrange the equations as necessary	Amber: A weightlifter lifts a 860 N barrel above his head to 2.2 metres in height. It takes him 10 seconds to do this. What is his power?	Amber: Work done (N) = Force (N) x distance (m) = 860 N x 2.2m = 1892 J	Amber

		Power (W) =Work done (J) / time taken (s) $$= 1892 \text{ J} / 10 \text{ s}$$ $$= 189.2 \text{ W}$$	
	Green: Another weightlifter has a power of 200 W. He lifts the same weight in 5 seconds. How high does he lift the weight?	I can't do this green bit yet	Red

Past paper questions

Even though science GCSE specifications are subject to change, sample or previous exam questions can still provide two key benefits. First, they give the students an insight into what they need to be able to do and feedback on how to get there. Second, they provide you with frequent opportunities to really understand what is expected of the students at the end of their studying which, in turn, can help you to understand the assessment criteria.

Using past paper questions well is important. This does not involve just giving your class a past paper to complete every now and then. It requires carefully planned activities to show students how to do better.

Here are some techniques you could try.

The improving carousel

Put students into groups of three and give each student a different past paper question on the same topic. They should complete the question in a period of time that you have defined for them (ideally each question should be worth the same number of marks). Once the time is up, they pass the question clockwise to the next student who has half the amount of time to improve on the first student's work using a different coloured pen to distinguish their work. The third student in the group then marks the answer using the mark scheme. Finally, the students should go through the completed questions and reflect on what they learnt through the feedback process.

Targeted questions

You can employ this exercise after the students have completed a checklist or base it on a gap analysis from a test or mock exam (this is a useful exercise where you analyse the performance of the students question-by-question to look for strengths and weaknesses). Organise students into groups based on them having similar areas that they need to develop and give them questions appropriate to each group's need.

The leader board

The leader board is best used in the run-up to exams when students are in full revision mode. Use one 10 mark(ish) question at the beginning of each lesson, which should link either to the previous lesson or to a lesson from some point in the last week (remember how useful spaced learning is here). Students should then either self or peer assess their answers. The teacher collects in the marks and keeps these on a leader board that builds over time. My experience of this activity is that it almost always rewards those students who work hardest outside of the lesson rather than the most 'gifted' individuals, so it rarely causes any real issues provided the relationships in the classroom are right. However, if you are concerned about the feelings of low-ability students, then you could display just the top 10 in the class. This technique also generates data to support you in identifying students who may be falling behind.

DEVELOPING SCIENTIFIC SKILL

So far, much of the discussion and techniques in this book have focused on students' learning and the understanding of key words. With so much detail in the science curriculum, it is possible to lose sight of the need to develop the students' scientific skills. To develop 'deep learning' our students need a solid overview and a good quantity of knowledge if they are going to go on and make the links required to have a nuanced conceptual understanding of a topic.

Researchers at the University of California at Berkeley, who have looked into how to best develop science skill, found that it is often poorly taught and that what teachers need to do is to integrate scientific knowledge with real life problems. They suggest that we need to support students in 'developing a repertoire of ideas, adding new ideas from instruction, experience, or social interactions, sorting out these ideas in varied contexts, making connections amongst ideas at multiple levels of analysis, developing more and more nuanced criteria for evaluating ideas and formulating an increasingly linked set of views about any phenomenon'.[18]

---

18 M. C. Linn, 'The Knowledge Integration Perspective on Learning and Instruction', in R. K. Sawyer (ed.), *The Cambridge Handbook of the Learning Sciences* (New York: Cambridge University Press, 2006), pp. 243–264 at p. 243.

For us to do this, we must be far more explicit in the skills that we teach students both in our planning and in communicating our intentions with our classes. The best way to go about this is to map out how you are going to develop the essential scientific skills over time, which is best done with your colleagues. Skill development needs planned distributed practice in order for students to move forward. Deep learning therefore takes time and requires a concerted effort by both the teacher and the students. John Hattie and Gregory Yates's *Visible Learning and the Science of How We Learn* (2013) is a really good starting point for understanding how to do this in the classroom.

The key skills that need to be developed include:

- The scientific method (including the teaching of the jargon, e.g. variables, precise, accurate, reliable, valid, proportionality).

- Problem solving and being curious about what we do not know.

- How to share data concisely and accurately in charts and tables.

- Forming conclusions and evaluating data and reasoning.

- Scientific numeracy.

- Practical skills.

- Evaluating science in the media.

- Making and using models.

It's a worthwhile task to think about your teaching and rate out of 10 how well you develop these skills to make sure that you are delivering on each one.

Developing skills is a difficult task, mainly because of the amount of time it takes to cultivate them. The following chapters look into how you can move this forward by:

- Focusing on literacy in every lesson so that students can comprehend text and can fluently articulate their understanding (Chapter 4).

- Looking at your curriculum and how skills are developed over students' secondary studies (Chapter 5).

- Planning in the long, medium and short term to ensure students can grasp challenging topics (Chapter 6).

- Examining how practical work and data is used in lessons to promote deeper learning (Chapter 7).

- Using assessment as a tool to help students understand how to develop their scientific skills (Chapter 8).

Most importantly, the teaching of scientific skills cannot be simply an add-on. It has to be integrated into lessons and concepts so that it is grounded in knowledge. For me, the most important thing here is to maintain curious students – ones who ask questions that you've never really thought about; the kind of students who walk into my class asking questions about *Interstellar* (which they'd just watched) or what on earth the Higgs boson is. Keep your classroom open, exciting and flexible and you'll get this kind of inquisitiveness. The other really essential outcome is having students leave me in Year 11 being sufficiently scientifically literate to become active members of society, whether or not they are going to continue with the study of science.

Chapter 4
THE IMPORTANCE OF EXPLICITLY TEACHING LITERACY

'EVOLUTION IS LIKE THAT WE'RE THE CHILDREN OF CHIMPS OR SOMETHING'

Literacy isn't just for English teachers. If your science students can't comprehend texts or properly understand what a question is asking for, how are they going to pass the exam? Literacy development is not simply about how you support those with weaker reading skills, but how you develop all students so they can understand technical scientific literature. Our expectations are crucial here: all students should and must be able to understand challenging scientific work.

First, and most obviously, there is the matter of the sheer amount of key vocabulary that we are required to teach. There are thousands of new scientific terms in Key Stage 3 alone, a number which almost rivals the word count learnt in the French classroom (don't forget, science is a new language for students too). One of the main issues with teaching literacy in science is that students arrive with a layman's understanding of many scientific words which they have picked up outside the classroom. We often struggle to teach the concepts of mass, weight, energy and power as students have become hardwired by understandings they have gained from the popular media (and sometimes from primary school) that these terms are interchangeable.

Whilst it's possible to understand a concept without in-depth knowledge of the key words, they have an important role in making ideas more concrete to students and, later, supporting links between different ideas in their mind.[1] Key words are not just jargon; they are a way of solidifying ideas, naming them and making them more readily retrievable. Learning the language of science is a major part of every science lesson and this learning has to be plotted out carefully. It is crucial to make the key words so familiar to students that they can use them with the fluidity that comes from repetition.

For them to be able to do this, first of all you need to teach the key words clearly and with specific emphasis. Make sure you vary the tone and pace of your voice in order that the key word hangs in the air and then repeat them to the class again and again. To ensure that this is not just rote repetition, employ some of the methods described in this chapter but also question individuals frequently, insisting on their correct use in student responses. The teacher has to be the model of language use for their students, so think about how frequently you use the key words yourself and in what circumstances. But also consider the non-scientific language that provides the context as this can also present real issues for students. I remember one of my Year 11 classes coming out of a science exam flummoxed by the use of the word 'fluctuate' in one of the questions; their lack of understanding of the contextual literacy around the science had been a complete barrier to them understanding what the question was asking for.

Literacy, like all skills, is improved through practice combined with feedback. To really develop students' literacy you will need to find quick methods to check for understanding, which is what many of the following ideas

---

1   For example, when I was at school studying science (I may show my age now), we never used the key words 'reliable' and 'valid' when discussing experimental results. We just commented on their strengths and areas which would require further study/observation, but we did not understand the topic any less than if we had used the key words (ditto for independent variable, dependent variable and control variable). However, some exam boards now expect these terms to be taught explicitly.

encompass. You also need to make feedback on literacy integral to your assessment by marking specific sections of their work for its clarity. In GCSE exams, spelling, punctuation and grammar are judged in the '6 mark' questions; it is therefore advisable for students to practise these skills from Year 7 onwards.

A good exercise with your colleagues would be to review your exam board's expectations for how literacy is marked (most boards supply exemplars which you can mark blind before reviewing the marks that the exam board would have awarded), and then think about how you can provide students with opportunities to practise these skills. Use the guidelines in your marking to help you with these extended sections and ensure that you provide enough time for the students to redraft their answers and act upon your feedback (see Chapter 8).

Vocabulary ninja

One way of ensuring that students understand all of the key words, and remember them, is to play vocabulary ninja. Essentially, this is just a key word test that you do each week but the name acts to engage the students – they can only become a vocabulary ninja if they have achieved 9 or more out of 10 every week (alter the ninja pass mark according to the attainment level of your class). You might want to keep a leader board or you can award a ninja of the week/half-term. This works much better than simply doing a key word test each week, and you can personalise the idea to the students in your class (e.g. Strictly Come Science).

Who am I?

Each question is a clue to a key word in this activity. It is good as a starter, especially on a Monday morning.

Missing vowels

Missing vowels is an effective key word reminder to help keep them in the students' working memory. Display key words on the whiteboard with the vowels missing. Give the students a time limit in which to work out as many as they can, and stop when the first student gets them all. This can be harder than you (or they) think. Then get them to write definitions for each term in their own words.

Here are some examples:

D_L_T_

_LK_L_

HYDR_CHL_R_C _C_D

_ND_C_T_R

_N_V_RS_L _ND_C_T_R

C_NC_NTR_T_D

S_LPH_R_C _C_D

N_ _TR_L_S_T_ _N

N_ _TR_L

B_S_

C_RR_S_V_

N_TR_C _C_D^2

Call my bluff

Call my bluff is great for learning key words, especially when the definition is very precise.

For example, is half-life:

- The time it takes for you lot to reduce my life expectancy by exactly one half?

- The time it takes for half of the radioactive isotopes to decay?

- Half the time it takes for all of the radioactive isotopes to decay?

- Half the time it takes for half of the radioactive isotopes to decay?

---

2   Answers: dilute, alkali, hydrochloric acid, indicator, universal indicator, concentrated, sulphuric acid, neutralisation, neutral, base, corrosive, nitric acid.

Literacy exit passes

This is a nice plenary. Provide the students with the key words covered in the lesson which they should construct into a story. It doesn't have to be scientific in nature but ideally it should be done imaginatively to help make the key words memorable.

Here is an example using the key words: ionic, bond, negative, positive, metal, non-metal, high melting point, aqueous, conduct and electricity.

> Chlorine was a *negative* boy until the day he came across Sodium, the *positive* girl in the *Aqueous* Club. When they met there was a spark and they formed an instant *ionic bond*. Ionic bonds only form when a *metal* girl and a *non-metal* boy meet. The temperature rose in the club as they have a *high melting point* and because it was the Aqueous Club it was so electrifying they could *conduct electricity*.

Writing breaks

In order to maximise the effectiveness of extended writing opportunities, plan writing breaks into your lesson. This is a five to seven minute break (for you as well!) during which the students have to summarise the lesson so far in their own words. Plan some prompts for them, if needed, by providing lists of the key words required in their answers or a checklist of what must be included to be awarded, say, bronze, silver or gold for the task. This also gives you the chance to go round, have a look at what the students have written and, more importantly, at what they have forgotten, thereby letting you know what they haven't yet got.

For example:

Task: Review what we have learnt today on the life cycle of a star (10 minutes).

- Bronze: Use at least eight key words to describe the life cycle of a normal sized star. Explain at least two of the stages.

- Silver: Use at least twelve key words to describe the life cycle of both a normal sized and a massive star. Explain at least four of the stages.

- Gold: As silver, plus explain in detail at least six of the stages in the life cycle.

Online literacy resources

There are lots of online resources that can be great fun and educational too:

- https://bubbl.us – A brainstorming website which is good for homework.

- www.cram.com – This website creates flashcards to help students learn key words. It's never too early to teach someone to revise!

- https://quizlet.com – Quizlet produces quizzes and flashcards for key word learning.

- www.memrise.com – Memrise provides pre-made quizzes covering most topics and exam boards to support key word acquisition.

- www.mmlsoft.com/index.php/products/tarsia – Formulator Tarsia software generates triangular puzzles for key words and definitions.

- http://splasho.com/upgoer5 – The Up-Goer Five Text Editor is a great homework activity: you simply provide students with a question that they have to answer. The challenge is that they can only use the 1,000 most commonly used words in their response.

  For example, here is how you might explain homeostasis:

  > Homeostasis is keeping everything almost the same in the body. One part of this is keeping the body 37. You do this by making your blood either get closer or further away from the skin. You also can do that thing that makes you smell or you can move lots ever so slightly. The hairs on your skin can stick up to get a layer of hot air to keep you warm like or lie down.

  Clearly, this will not help with writing excellent answers for exams, but the process of doing it is so painful that the students will really have to think about it and, therefore, hopefully remember it in the process. This task forces the students to reflect on what they have learnt in class and, rather than learning the definitions by rote, to instead think of their own way of explaining the same idea.

- www.tripticoplus.com – Triptico offers a number of great tools including:

  » A task generator.

  » A task timer which counts down the time that students should spend on each task. This is really helpful when the students are working in groups and have to be time conscious.

  » Image bingo: the students choose five words from a list of key words on the topic but instead of providing definitions as in typical bingo you show them an image which represents the word. Continue showing images until someone has all five of their key words struck off and calls 'Bingo!' This is a useful five minute plenary for a lesson which has introduced a number of new key words.

» What's the question? The students use clues to work out the question to the answer you have provided. This is a good starter, especially if you've used a question as the title of your lesson.

• www.wordle.net – Create word walls for topics using Wordle and display them during your writing tasks.

• http://wordsmith.org/anagram – Wordsmith produces anagrams of key words that can be used as a really quick starter (insist that the students provide a definition once they have unscrambled the anagrams).

LONGER ANSWERS

Quite rightly, longer answers (those dreaded 6 mark questions) have returned to science exams and, as such, teachers from Year 7 upwards have to support students to write at length about complex topics. To do this, the students will need to have a deeper understanding of a topic, rather than just being able to parrot out the key words you've taught them. Here are some methods with which you can scaffold longer answers for them.

Graphic organisers

Graphic organisers can be useful to help the students link ideas from key words together and make sense of the information themselves. Whilst this may slow down the 'delivery' of the topic, it will increase their level of mastery. A graphic organiser is also really useful for independent study as it forces the students not just to copy and paste information from primary sources.

Graphic organisers usually start with a basic mind-map, which is valuable for collecting students' existing background knowledge about a subject. Some imaginative ways to use mind-maps include:

- As a large classroom display. Ask the students to add the new information they've learnt to it lesson by lesson until the mind-map is complete by the end of the unit.

- Carousel mind-mapping. Have six different topic brainstorms on tables of six students so that each student starts with one mind-map on one topic. Give the learners two minutes with each mind-map to record everything they know before moving the mind-maps clockwise around the group. Continue until each student has their original mind-map returned to them.

The beehive

The beehive organiser is useful when there is a hierarchy of ideas that students build up from the base. Use a box below the beehive to summarise the content they have added above in their own words.

Title: How the Earth's atmosphere evolved

 79% nitrogen, 20% oxygen, 1% other

Animals that respire evolved, using up more oxygen.

Plants evolved which photosynthesise, using up carbon dioxide and producing oxygen.

The water vapour condensed and formed the oceans. Some carbon dioxide was dissolved in the oceans.

The Earth was covered in volcanoes which released carbon dioxide. No oxygen was present. There was also water vapour and methane.

Summary:
The Earth was formed 4.5 billion years ago. Originally the atmosphere was mainly made of carbon dioxide which was released from volcanoes. Over time water vapour originally in the air condensed to form the oceans, dissolving some of the carbon dioxide. Plants evolved and during photosynthesis reacted carbon dioxide with water to form glucose and oxygen. These days the atmosphere is mainly nitrogen and oxygen.

The Cornell note taking system

The Cornell note taking system is useful for older students, particularly those in Year 11 or the sixth form, as it will help to prepare them for studying at university and the copious note taking that higher education often requires. The benefit of this method is that it encourages the students to think about their notes rather than just taking them.

Notes are recorded on the main section of the page and the margins of the paper are used to record any key words that are relevant to the lesson as it progresses. As a plenary, give the students 10 minutes to complete a summary of the lesson at the bottom of the page. Students can use these notes during revision by covering up all but the margins of the paper and recalling what would be in the notes section, before checking to see what they may be missing.

Cue column (useful for key words and definitions)	Is there life out there?
SETI – search for extraterrestrial intelligence. Monitoring electromagnetic radiation from space using telescopes.	In this area, students do the following: • Take notes in bullet form only, with no copying. • Note questions they have. • Reflect on their learning – what do they now know.
Summary box – 10 minutes after note taking, bring the class back to the summary box. Get them to cover the area above and summarise what they think they have learnt.	

The Frayer model

The Frayer model is a graphic organiser which helps students to chunk a topic down into its constituent parts. This can be completed before they attempt a longer answer question.

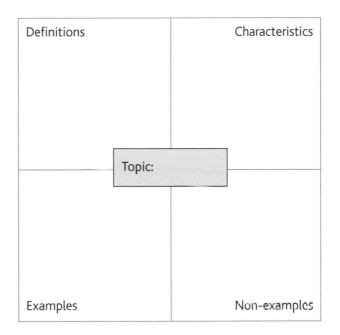

Hexagons

This idea comes from SOLO taxonomy[3] which splits learning into five levels:

- Pre-structural – learners have simple ideas about a topic but no idea of how the ideas fit together (e.g. leaves are found on plants, living things are made of cells).

- Uni-structural – learners know some things about the topic and can make simple connections between the ideas (e.g. plants are made from plant cells which have a vacuole, cell wall and chloroplasts).

- Multi-structural – learners understand how different ideas from a topic come together but can't yet understand them as a whole (e.g. plant cells are different from animal cells because they do not have skeletons and they needs chloroplasts to make energy during photosynthesis).

- Relational – students can understand the topic as a whole and how all of the ideas fit together (e.g. what occurs in the mitochondria and chloroplasts and why respiration and photosynthesis occur simultaneously).

- Extended abstract – this is the stage where students have the ability to relate ideas to other areas and think beyond the current context (e.g. starting to ask questions such as, 'How did life exist before chloroplasts?' or 'Can we genetically engineer our cells to have chloroplasts?').

SOLO can be a useful tool with which to think about what you might expect the students to do to move themselves forward, and it can be a neat way of explaining to them how to go about developing their understanding in interesting and useful directions. The best way I've seen SOLO used is with hexagons. These cover all of the structural areas which means this task can

---

3   J. Biggs and K. Collis, *Evaluating the Quality of Learning: The SOLO Taxonomy* (New York: Academic Press, 1982).

link knowledge and skill development and is a great way of differentiating during revision lessons.

Use your technicians to help you with this – I have a folder filled with card hexagons kindly prepared by mine. Give a handful of hexagons (preferably laminated so they can be written on with non-permanent markers) to a group of three students. Start by getting them to write one key word on each hexagon from the topic studied (you can give them the words if you think it's necessary), and then get them to connect the hexagons into a shape and justify why the words have been linked in the way they have. Finally, ask the students to annotate the hexagons with what links them together and then ask for any questions that this activity has brought up for them.

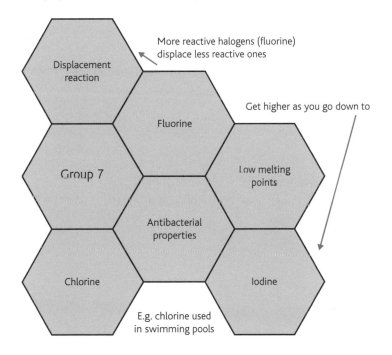

The BUS method for reading

The BUS method promotes the reading of questions carefully before moving on. Students should:

B – **Box** the command word.

U – **Underline** the key words in the question.

S – **Scribble** additional ideas and words that spring to mind from the question.

Reluctant readers will often skim questions and then start answering them nigh on immediately without having given the question any proper thought, whilst other students are often in a rush to get their work finished so they can have a rest with this method. It works best if the teacher models BUS to students a number of times before they have a go themselves.

Use their ears

Elephants are the largest animals that live on land. Elephants have difficulty keeping cool in comparison with smaller animals in the same environment. Explain why

Use ideas about energy transfer in your answer   [6 marks]

hot!   less surface area

The POW method for writing

Quite often, a student will have knowledge about a topic, or even go as far as understanding it well, yet they will still struggle to demonstrate this understanding concisely when presented with a piece of paper. This is a huge issue, particularly when students are preparing for exams, and even more so when they are stressed in an exam hall. A good technique to teach them is POW!

P — **Pause.** Stop for a second, put your pen down, pick up a highlighter and ensure you have fully understood the question.

O – **Organise.** Plan out your answer. This could be a flow chart of key words in the order you will use them, a table comparing and contrasting two ideas or simple bullet points which you subsequently number in the order you want to include them in your answer.

W – **Write well.** Use your plan to write your answer. At the same time, underline all of the key words you have used to ensure they are there (and are explained). Tick off each aspect of your plan after you have included it in your response. Re-read your work.

Take me down the river

Students often understand concepts but find it hard to explain them in the requisite level of detail. Often, when explaining a sequence of events, a typical answer will involve the first point in the sequence, perhaps one in the middle and a final point. To overcome this issue, I give students an image of a river (you could probably find another equally apposite analogy) with the same number of bends as there are points to the answer. As a class, we agree the start point of the river and the end point, then the students (in pairs) have to work out how to get down the river and what points need to be written on each bend. Challenge the students to find even more bends to add to their answer.

This example shows the starting points you may want to give students if you would like them to explain why a sprinter still breathes heavily even after the race has been completed.

The sprinter needs energy.

Oxygen reacts with lactic acid to make $CO_2 + H_2O$

A similar technique is to get students to work out the six degrees of separation between two statements. This is a take on the Six Degrees of Kevin Bacon (there is a lot about this on the Internet) which works on the premise that any two people on Earth are only six acquaintance links apart. In the classroom, you can give the students two statements and ask them to marry them up with six connecting statements (the rule is that they must use full sentences and can't use less than six). This can be used not only to help students extend their writing (as above) but also to challenge them to link different ideas in science (e.g. How could you link Marie Curie to the element argon?[4]).

MOVING LITERACY FORWARD

The development of literacy skills for our most able students is often overlooked in science lessons. In their review of science lessons, Ofsted noted that with regards to literacy 'some teachers had focused too narrowly on scientific literacy, such as the correct use of technical terminology' rather than developing more complex writing skills.[5] And it is often the highest achievers who are least likely to be properly catered for. The challenge for teachers is to find that golden area of 'desirable difficulty' where higher attaining students find the work challenging but manageable.

Scientific publications should be introduced and explained early on in a high attaining student's journey through science. How to go about understanding the language conventions in them needs to be modelled and the teacher should guide students through the publication(s), explaining the thinking

---

4   Here's one way: (1) Marie Curie worked with radioactivity. (2) One of the elements that she worked with was plutonium. (3) Plutonium is found in the periodic table. (4) Mendeleev created the periodic table by putting all of the elements in order of atomic size and grouped them by similar properties. (5) Mendeleev created the group 0 for the noble gases as they did not react with other elements. (6) Argon is a noble gas.

5   Ofsted, *Maintaining Curiosity: A Survey Into Science Education in Schools*. Ref: 130135 (2013). Available at: https://www.gov.uk/government/publications/maintaining-curiosity-a-survey-into-science-education-in-schools, p. 35.

behind the text, graphs, tables and thought processes that have led to the conclusions. We should be giving our highest attaining students as many opportunities to read for scientific meaning as possible.

You might be questioning why teaching numeracy is not included here. The reason is that whilst there are elements of numeracy that students must master in order to be good scientists, this will not be possible in every one of your lessons – unlike literacy, which is pivotal in more or less every lesson. A 'how to' guide on teaching numeracy in the science classroom can be found in Chapter 6.

Chapter 5
THE JOINED-UP SCIENCE CURRICULUM

'BUT WHAT DOES THIS HAVE TO DO WITH WHAT WE DID LAST WEEK?'

You may be sat reading this book as an NQT and thinking that a chapter on planning a curriculum can't possibly be aimed at you (at least not yet). However, thinking long term is one of the most important things you can do as a teacher because without thinking on this slightly grander scale it can be impossible to see the bigger picture. Neglecting to do this can be dangerous.

If you were to teach a unit on, say, the periodic table to a Year 7 group and you haven't thought about how this will develop into future topics, then it is likely that you will lose emphasis on why the periodic table is constructed in the way it is and focus instead on the elements and the history of its construction. In omitting this emphasis, the students will later struggle to understand how to apply the periodic table to predict properties and understand trends within the groups.

Simple mistakes can also be made if you don't plan at the medium and long term levels – an NQT I knew made the error of not realising he had finished teaching a topic and neglected to tell his class that they had to do a test in the next lesson. Needless to say, they were not happy when they showed up to a test they hadn't prepared for!

The national curriculum is fairly flexible these days at Key Stage 3, and this gives departments the opportunity to think about how best to deliver core scientific ideas so that students are ready to study for their GCSEs when they reach Key Stage 4. A great curriculum can help you to plan how to balance knowledge and skills development over time, as well as making sure that there are ample opportunities for students to be excited by and engaged with the subject. Beyond that, the curriculum should also be flexible enough to provide each teacher with the ability to tailor how they teach the class in front of them.

HOW TO COPE WITH THE CURRICULUM WHEN YOU ARE NOT A SUBJECT SPECIALIST (OR EVEN IF YOU ARE)

A whopping 35.5% of science teachers are biology specialists (in comparison with just 14.5% who have a physics specialism),[1] which means there is a lack of chemists and physicists in our schools and that a vast number of science teachers are teaching outside of their degree subject. The breadth of topics we are required to teach is pretty vast and very few science teachers have the extensive subject knowledge they need for the job at the beginning of their career. That said, there are benefits to teaching science as a single subject: not only does it allow you to get to know your students better but

---

1   Royal Society, *The UK's Science and Mathematics Teaching Workforce: A 'State of the Nation' Report 2007* (London: Royal Society, 2007). Available at: https://royalsociety.org/~/media/ Royal_Society_Content/education/policy/state-of-nation/SNR1_full_report.pdf, p. 26. House of Lords, *Science Teaching in Schools: Report with Evidence of the House of Lords Science and Technology Committee*. HL Paper 257 (London: The Stationery Office, 2006), p. 46.

it can also provide better continuity between the three disciplines and offer more opportunities to link different areas together for students.

There is a 'but' though. You will naturally be better at one subject depending on the degree you have, so the challenge will be to make sure that you are just as passionate about all three areas of science. Hopefully, there will be someone in your department who has a passion for your weak areas (mine was electricity to start with) and who can support you in ensuring that your students don't notice it's your weak spot. The issue though, as ever, is how you manage your time as you'll find that you are unlikely to want to spend a couple of hours learning how to make electricity exciting on a Thursday evening in the middle of term.

The reality is that, particularly at the beginning of your career, the best time to do this is in your holidays. Yes, after a long term of teaching you will still need to use a lot of your holiday to relax, sit on a beach, have a lie-in and watch *Homes under the Hammer*, but you will also need to spend some of your time keeping up with your subject. Before each half-term starts, review the topics taught, talk with members of your team, use science blogs and read science magazines. In addition, if you have a group studying towards taking a GCSE exam soon, make sure you go through the exam papers for the topic to get an idea of how they will be assessed and what you need to plan to get them there — subject knowledge is more than just science content. If necessary, brush up on your wider subject knowledge: just reading the textbook will not be enough to open students' eyes to how amazing science is. In addition, pedagogy moves forward and it is just as important to avoid getting stuck in a rut with the way you teach, so you need to keep on top of that as well.

KEEPING UP WITH YOUR SUBJECT KNOWLEDGE

Subject knowledge is not an area of concern solely for new teachers. Science is a fast moving subject and we all need to keep up to date with changes and new discoveries in the field. Here are some books and blogs that will help you to keep up with developments in the area:

- *New Scientist* – Consider having a subscription for your school, for both staff and students.

- *The Oxford Book of Modern Science Writing* edited by Richard Dawkins (2009) – I read this book as an NQT and it gives concise descriptions of the major ideas in science.

- www.thenakedscientists.com – Really useful for finding out extra interesting bits of information to build your subject knowledge in your not-so-strong areas.

- www.popsci.com – Provides up-to-date scientific news stories which can work as perfect hooks for students. It's great for printing things off and posting them on your classroom door to help the students understand that science is continually moving forward.

- http://wefuckinglovescience.tumblr.com – Great ideas, although you will need to edit some of the language first.

- www.wired.com/category/science – Picks out some of the best new scientific developments.

THE CHARACTERISTICS OF A GOOD SCIENCE CURRICULUM

In this section we will look at the main characteristics of the science curriculum: hinges, spacing it out and meeting the needs of all.

CHARACTERISTIC 1: THE HINGES

The starting point of any curriculum must be the key concepts that the students will have to grasp in order to have a decent grounding in the subject. These are sometimes referred to as 'threshold concepts', which reinforces the idea that within all areas of study there are concepts so central to the subject that understanding them can result in thinking about a topic in a previously inaccessible way. By streamlining our curriculum to these big ideas and planning around them, we enable the students to make much bigger leaps in their understanding.

There are also topics that experienced science teachers will immediately be able to identify as areas that students struggle with. For example, if evolution is taught in too little detail early on in their school life, older students can struggle to understand complex ideas such as protein synthesis and other areas at the molecular level. The failure to understand properly what a particle is can result in later difficulties with comprehending what titration is about or what intermolecular forces are. In physics, calculations can cause problems for students well into A level and electricity still baffles most GCSE students.

We have a tendency to shy away from these troublesome ideas, boiling them down until they are so simple that we don't challenge students to properly understand them. This is one of the reasons that when students begin A level science, they are often told by their teacher to forget all that they have been taught at GCSE. Students (and sometimes teachers) can tend towards the assumption that if a lesson went well and was easy to learn then they have been successful. In reality, research has shown that so called 'desirable difficulties' are more likely to result in memorable learning and, as such, it is important that these are given the time they deserve.[2]

Listed in the table below are the fundamental ideas that make up science – those which form the key hinge points which students need to grasp. I have left out a number of topics – we can get too caught up in the idea that *every* area of science is important and this can sometimes get in the way of a much better understanding of the core principles. Underpinning all of these areas are the scientific skills that are critical if students are going to understand the true nature of science.

Biology	Chemistry	Physics
Evolution and genetics	*Particles* and states of matter	*Energy* and waves
Cells and DNA	Bonding and rates of reaction	Forces and motion

---

2   See E. L. Bjork and R. A. Bjork, 'Making Things Hard on Yourself, But in a Good Way: Creating Desirable Difficulties to Enhance Learning', in M. A. Gernsbacher, R. W. Pew, L. M. Hough and J. R. Pomerantz (eds), *Psychology and the Real World: Essays Illustrating Fundamental Contributions to Society* (New York: Worth Publishers, 2011), pp. 56–64. For a worthwhile overview of desirable difficulties have a look at P. C. Brown, H. L. Roediger and M. A. McDaniel, *Making it Stick: The Science of Successful Learning* (Cambridge, MA: Harvard University Press, 2014).

Independence – photosynthesis, respiration, ecology and the importance of plants	Making predictions – the periodic table	Space
The body – hormones, organ systems, disease	Earth science, hydrocarbons and energy	Electricity

The idea of teaching core ideas really well has been fully embraced by the education system in Singapore, which is widely known for teaching science better than anywhere else in the world (at least according to PISA test results). Theirs is a 'teach less, learn more' curriculum that results in students focusing on the big ideas and on developing scientific skills rather than getting bogged down in unnecessary levels of detail. The emphasis is on depth of knowledge and skill development alongside a rigorous checking of understanding, with the expectation that no student should fall behind. As Daniel Willingham observes, 'Practice makes perfect – but only if you practice beyond the point of perfection,'[3] and this can only really happen if it has been catered for in your long term planning.

Of course, the impact of this is that if you teach some topics really well, there is a trade-off with the number of topics that you can get around to teaching. You may notice that topics such as electromagnetism and new materials are missing from the table above. This is because, whilst they are important, they do not make a fundamental difference to a student's grasp of core scientific ideas. (Their omission here does not exclude them,

---

3   D. T. Willingham, 'Practice Makes Perfect – But Only If You Practice Beyond the Point of Perfection', *American Educator* (spring 2004). Available at: http://www.aft.org/periodical/american-educator/spring-2004/ask-cognitive-scientist.

however, from being taught to those students who are ready to learn about wider areas of science.)

The three topics highlighted in italics (evolution, particles and energy) are the key hinges for students to understand as they underpin so many of the subjects we teach. In much the same way that an English teacher would constantly teach grammar skills or spelling, these topics need to reinforce how we go about our teaching.

EVOLUTION

Nothing in biology makes sense except in the light of evolution.

Theodosius Dobzhansky[4]

The theory of evolution is critical for understanding biology – from why interdependence is so important to why we have villi in our small intestine. Often our teaching of evolution comprises a few lessons in a Key Stage 3 scheme of work, but in terms of mastering biology it is essential that it is taught in depth to ensure that the understanding of big concepts – such as genetic disease, protein synthesis and interdependence – is properly under-pinned at Key Stages 4 and 5.

This is one area where, if you can manage it, you should try to drum up support from other departments in your school: history, RE and English can all have a role to play in developing the hinge. Ask history teachers to further embed this knowledge by teaching what the Victorians thought about Darwin or ask them to study the emergence of the idea of 'intelligent design' in the United States. The RE department can be covering the debate on

---

4   T. Dobzhansky, 'Nothing in Biology Makes Sense Except in the Light of Evolution', *American Biology Teacher* 35 (1973): 125–129.

evolution, whilst we cover the scientific facts. The University of Cambridge's Darwin Correspondence Project produces some brilliant resources that are well worth any science teacher finding out about and using.[5]

Many of the ideas in evolution are intuitive for students, but misconceptions and an inability to link concepts can be a barrier. Before you start this topic, ensure the students have a basic foundation in the structure of cells, adaptations, genes and environments. A pre-assessment (see Chapter 8) may be useful here. This scheme is a lengthy one but it is a good example of where a 'mastery' curriculum can be really useful.

Foundation activities – ideas to get started with

Play the 'Right Here, Right Now' video by Fat Boy Slim which covers 350 billion years of evolutionary biology in 3 minutes and 40 seconds flat and features men in gorilla suits dancing to 1990s house music. Ask your students to discuss the video. What do they think of it? What were the key points? What did they think was incorrect about the rather impressionistic story the video tells?

Then ask the students to discuss what an adaptation is (something that should come naturally) and how many examples of this they can think of (use a think–pair–share here – see Chapter 2). Reflect with them on why these adaptations are so important and what would be the likely outcome if organisms were not adapted to their environment.

At the end of your introductory lesson use a 'question hat' to give the students an opportunity to ask you something – hand out sticky notes and get them to write down one thing they are interested in finding out about,

---

5   See http://www.darwinproject.ac.uk/schools-resources.

which they then place in your hat as an exit pass before they leave the classroom. They will often ask the most incredibly imaginative questions which you can then use as a great springboard for discussion points in the next lesson(s). Their questions will also allow you to spot any misconceptions they may have.

You can bring the history of evolution alive through the story of Darwin. There are some brilliant clips to be found on YouTube to help you do this. For example:

- He was born into a family of free thinkers who taught him to think for himself and question what he learnt. His grandfather wrote, 'would it be too bold to imagine that all warm-blooded animals have arisen from one living filament ...?'[6]

- He was not a good medic and dropped out of medical school.

- He spent five years on HMS *Beagle* travelling the world, spending most of his time at sea reading and thinking about what he had collected on his travels.

- The evidence he collected from the Galapagos Islands is the neatest example of his theory.

- Both *On the Origin of Species* (1859) and *The Descent of Man* (1871) were met with substantial controversy on publication.

- He warned others about the dangers of in-breeding based on his own experience of having children with his cousin.

Provide groups of students with different pieces of classic evidence on evolution, ranging from how horse legs evolved through the fusion of multi-toed feet into single-toed hoofs to the adaptation of finches' beaks to be

---

6   See E. Darwin, *Zoonomia; or the Laws of Organic Life*, Vol. 1 (London: Johnson, 1794), p. 397.

the perfect size for the seeds found on their island homes in the Galapagos. Get the groups to come up with their own ideas on what they think must be happening in each of the examples before you introduce the theory of evolution, providing question prompts to help them develop their understanding (e.g. What has happened to the legs of the horses over time? Why do you think this is? What would be the advantage of this? What is the correlation between beak size and size of seed? Why could this be useful? Why would there be more of these kind of finches than other ones?).

Use the analogy of the evolution of a tree (as opposed to a ladder) as this is one of the clearest ways for students to understand the idea of the common ancestor. The most widespread misconception that students have about evolution is that we have evolved directly from chimpanzees (like a ladder), so a tree analogy helps them to understand that humans and chimpanzees share an ancestry and are more like cousins than father and son. Diagrams will help the students to visualise this idea. Make sure you go over this again and again – one way to do this would be to get students to download the TimeTree app (it's free) or use the website (www.timetree.org) to look into the evolutionary tree in more detail.

Create an evolution timeline along one wall of your class over the space of the unit (you can keep coming back to it throughout the year). Include the development of bacteria, the evolutionary 'big bang' in the Cambrian period, plants and dinosaurs, and go all the way up to the modern day (and perhaps into the future).

Techniques for taking it further

Once your students have fully grasped the general idea of evolution, the next stage is to make sure they understand the evolutionary process on something more than a superficial level. We can do this by making sure the students get an opportunity to engage with the data which supports the theory.

Here are some ideas to develop students' understanding of evolution:

- Tweezer food challenge. Give the students differently shaped tweezers and a Petri dish containing seeds to model the way in which finches' beaks evolved into different shapes to adapt to differing environmental constraints. Start with one set of tweezers and give the students 30 seconds to 'eat' as many seeds as possible by picking them up and moving them to the 'stomach' (which can be drawn on a sheet of paper). Then change the kinds of seeds you use to illustrate a new environmental challenge. It helps if these seeds can't be picked up by the old set of tweezers so you then introduce a new set with which the students are able to pick them up. This models the reactive and adaptive process of evolution in response to environmental factors in a way that is easily memorable.

- Red and green 'worms'. Take your students outside and equip them with a quadrat and toothpicks that are coloured either red or green (about 20 each per student pair works well). Ask them to place the quadrat on the grass and to distribute the toothpicks within the quadrat. They then have 10 seconds to act as 'birds' picking up the toothpicks and must compare the number of red or green worms they've each gathered. Back in the classroom, pose the question: what would happen if the grass suddenly turned red?

- Recreate some of Darwin's experiments from the *Beagle* (see the Darwin Project website for activity packs of information to use as stimuli, such as Darwin's letters, and also explanations of how to replicate the experiments in the classroom).

- A wordless essay (see Chapter 3) would work well here to summarise what the students have learnt from the experiments and to get them to link the practicals they have undertaken to the theory they are studying.

- A nice homework activity to examine why there are gaps in the fossil record is to ask the students to research their own family trees prior to the lesson. Most will end up with gaps and this can be used to illustrate how this happens in terms of our knowledge of each stage of evolution. (You might also use this when you cover genetics and link the two together.)

Techniques for developing scientific understanding

To help students think scientifically they must do something with the information you have taught them. Here are some ideas:

- Move forward from Darwin through to Wilkins, Franklin, Watson and Crick, introducing the concept of mutations as the mechanism for evolution.

- Discuss how, on occasion, random mutations in DNA can result in the evolution of a species by allowing that species to be better equipped to survive in their environment (whilst at the same time ensuring that students understand that most mutations have negative or neutral repercussions). Students love the story of how the FOXP2 gene allowed humans to develop speech and language.[7] On the flip side, look at how genetic diseases can evolve and still remain in genetic lineages.

- Study the pentadactyl limb. A good group work activity is to give each member of the group a different example of a pentadactyl limb (e.g. bat, horse, whale, human, bird) and get them to hypothesise how its evolution would have helped that animal to survive, before coming together as a class to compare their structures.

- Study the evolution of the Homo genus and why other species of Homo are now extinct.

- Consider which other species of animals whales are most closely related to.[8]

- Look at the example of cichlid jaws in Lake Tanganyika. The cichlids have evolved to eat the scales of other fish – some of the fish have a jaw that twists slightly to the right and others slightly to the left. Of the two forms, it pays to be part of the population that has the rarer jaw because this means the cichlid can surprise their prey (which have learnt which way the fish attack) and so get their dinner easier. One year the rarer form may be a jaw which bends to the left, then natural selection will work in its favour until it is the most popular kind of fish. Then the right sided fish will do better, and hence there is a continual cycle in the lake year on year.

---

7 In 1990, the KE family were the focus of a ground-breaking study on human evolution. Approximately half of the family suffered from severe difficulties in speaking and it was theorised that it was due to a single mutation in a dominant gene. Researchers located the gene and compared it to the DNA of close human ancestors and suggested that the gene might have been one of the most important factors in the evolution of language (and thus the evolution of humans). See W. Enard, M. Przeworski, S. E. Fisher, C. S. L. Lai, V. Wiebe, T. Kitano, A. P. Monaco and S. Pääbo, 'Molecular Evolution of FOXP2, A Gene Involved in Speech and Language', Nature 418 (2002): 869–872.

8 It's hippos! Use this piece of information to trace how evolution moves in mysterious ways.

- Consider whether elephants are evolving to be tuskless to avoid poaching.[9]

Techniques for developing mastery

Evolution is the kind of topic that it can be really useful to 'over teach' if you can. By this I mean really go to town on different examples of evolution and, at each opportunity that arises for the students to practise their understanding, go back to the key concepts of the topic. By the mastery stage, the students should be able to transfer and use their knowledge in new situations. There are some excellent ethical and philosophical areas that you could debate with the students:

- Should we put so much money into saving giant pandas?

- What would happen if a blight occurred that killed off all of our agricultural plants?

- Will humans eventually become a mono race (i.e. no distinction in ethnicity)?

- What would happen to human evolution if 'designer babies' became legal?

- Why does the flu vaccine last only for one year?

- How can fears be inherited (the science of epigenetics)?

---

9   They are – a great example of evolution in action. See R. Gray, 'Why Elephants Are Not So Long in the Tusk', *The Telegraph* (20 January 2008). Available at: http://www.telegraph.co.uk/news/science/science-news/3322455/Why-elephants-are-not-so-long-in-the-tusk.html.

PARTICLES

Those who are not shocked when they first come across quantum theory cannot possibly have understood it.

Niels Bohr[10]

Chemistry is made up of two main areas: the material world and the invisible particle world of atoms and molecules. Students can naturally identify with and comprehend the material world, but to understand it properly they have to understand particle theory and this can be very difficult indeed. That said, it can be fairly easy to make this topic fascinating to students as there are some brilliant facts in this area. For example, the human body is made up of such an awe inspiringly large amount of atoms that it has been theorised that up to 20 billion of the body's atoms during a lifetime would have once been part of William Shakespeare![11]

The massive challenge when studying particles is how we make something that the students can't see (and which embodies the most abstract of ideas) understandable, especially to Year 7s. It is not just the fact that we can't see them that causes the problem, but that the particle size itself is very difficult to get your head around (approximately 10,000,000 atoms would fit into the full stop at the end of this sentence).

Another consideration here is literacy related as there are so many words that students can tend to regard as interchangeable: atom, molecule, particle, proton, ion, compound. Right from the start, clarity on the exact meaning of these words is essential if the students are not to confuse them later on. One of the main issues students have with balancing equations is

---

10 N. Bohr, *The Philosophical Writings of Niels Bohr*, Vol. 2: *Essays 1932–1957 on Atomic Physics and Human Knowledge* (Woodbridge, CT: Ox Bow Press, 1987).

11 See http://www.jupiterscientific.org/review/shnecal.html.

understanding what a molecule is and why hydrogen or chlorine have to be diatomic.[12] You need to introduce the idea of a molecule at the right time – that is, once they understand the periodic table and how elements can combine into compounds. Most textbooks introduce elements, compounds and molecules at the same time (right at the beginning), but my experience is that the understanding of molecules is often lost when it is taught in this way and that this method adds a layer of complexity that isn't really needed when it's first taught. Only when the students have developed the required background knowledge should molecules be taught about as a specific type of compound which is made up of only one element.

So, be careful when teaching students about particles. Correctly identifying elements, compounds and mixtures in particle diagrams does not mean that the students truly get it, but only that they have learnt what the diagrams should look like and can regurgitate this information. To avoid this, interweave the ideas throughout your teaching of chemistry and physics. Particles are a complex topic that can't be taught in one simple step – they require careful consideration and revisiting across the five year learning journey. Many students will not be able to understand the abstract nature of this topic in Year 7 and will need a number of opportunities to deepen their understanding throughout Key Stage 3 in order to ensure they have enough coverage of the subject by the time they take their GCSEs.

---

12 Students need to be aware of the elements that commonly form diatomic molecules. Hydrogen, oxygen and group 7 are required at this level (for most of Key Stage 3 and 4 it is very difficult to explain to them why these elements, and not others, without a huge diversion). Have definitions and diagrams of elements, compounds, mixtures and molecules (including all the molecules that they need to know) up on your wall and remind the students to check it whenever you go through balancing equations.

Common misconceptions about particles that students bring with them in Year 7 are:

- Substances *contain* particles rather than being *made from* particles (e.g. a glass of water contains particles rather than being made up of particles of water).

- Chemical change is the observation of a reaction rather than the production of new particles.

- Particles are the size of a grain of sand!

- Particles are static and do not change.

Foundation activities – ideas to get started with

At Key Stage 2, students are likely to have been taught about changes of state and irreversible changes, states of matter and the idea of the particle as the basic building block. When beginning the study of this area at Key Stage 3 you should only ever refer to particles – do not complicate things by using any other word (e.g. atoms or molecules) as these can be taught explicitly later on, especially once the periodic table has been taught.

As discussed in Chapter 1, start with a historical overview of how our ideas in this area have developed. You can begin with the stories of Democritus and Dalton, but perhaps go further and discuss Marie Curie and Becquerel and how their lack of understanding meant they failed to see how the invisible radiation was affecting them.

Employ analogies to explain the scale of particles,[13] but beware of going overboard with PowerPoint slides of coloured balls as these don't really help to make the topic real. Instead, the best starting point is to undertake many small experiments and introduce ideas from real experiences to bring it to life:

- Get the students to separate substances into solids, liquids and gases, including looking at non-Newtonian fluids like custard (there are some brilliant videos on YouTube). Provide plastic bottles filled with water, sand and air – how do they differ when squeezed? What does this tell us?

- Provide illustrations/demonstrations of particle theory from real life such as crisp packets expanding on a plane and doors not fitting in the same way in winter and summer. Get the students to think–pair–share what is happening.

- Ask the students to put marshmallows in a vacuum pump and observe what happens.

- Attach drawing pins to metal rods with Vaseline, then ask the students to heat the metal rod around 10 cm from the drawing pins and observe what happens.

- Collapse a plastic bottle by submerging it into hot water up to the neck. After a few minutes, screw the cap on and plunge the bottle into cold water. What happens?

- Get the students to role play water in different states!

---

13 Check out Jon Bergmann's Ted-Ed video for some good examples: 'Just How Small Is An Atom?', *TED* (April). Available at: http://www.ted.com/talks/just_how_small_is_an_atom.

Bring the worlds of the particle and the real world together as many times as you possibly can. In doing so, you will be helping your students to learn the fundamental ideas of life!

Techniques for taking it further

Before moving on, a nice way to ensure the students fully understand things is to show them the video of *Apollo 15* Commander David Scott simultaneously dropping a feather and a hammer on the moon and discussing why they land at the same time.[14]

The next step to take to embed understanding is to teach diffusion really, really well. Diffusion is often taught quite poorly at Year 7 and, as a result, many students still don't fully understand it at Year 11 when they need to be able to apply it to a number of situations, from diffusion of oxygen and carbon dioxide in the circulatory system to rates of reaction. Suggested activities to start with in Key Stage 3 include:

* A carousel of diffusion mini practicals. Using their prior knowledge of particles, ask the students to write down reasons for what they think is happening in a quick lab report (for more ideas on practicals see Chapter 7):
  » Spray some perfume. What happens?
  » Add food colouring to water (or potassium permanganate) and write down observations after 5, 10 and 15 seconds.
  » Put gobstoppers in water and record what happens.

---

14 See https://www.youtube.com/watch?v=-4_rceVPVSY. In this experiment, Scott was testing Galileo's hypothesis that all objects, regardless of mass, drop at the same speed and that the difference that we observe on Earth is due to air resistance.

- Demonstrate the bromine experiment. Put bromine in one jar and air in another separated by a sheet of paper. Remove the paper, observe the bromine diffuse around both jars and discuss with the students what they think has happened. Most of them will conclude that the air jar is empty and that the bromine is simply filling up the entire jar. Use this as an opportunity for a think–pair–share activity, then explain the science. Use a smoke cell in a microscope (ideally connected to an interactive whiteboard) and introduce the idea of particles moving from a high to a low concentration to clarify the idea of constant random motion.

In another lesson, do the bromine work again by role playing the experiment and getting the students to act out being particles. For example, you could split the class into two: half are bromine molecules (get them to wear their blazers) and the other half are the various air molecules (not wearing blazers). Send them to opposite sides of the classroom. Get a few students to hold hands in a row to act as the paper keeping the two groups of students apart. Ask the students to randomly move about in their halves of the room (gently knocking into one another, the walls and the 'paper'). Remove the paper students, allowing the bromine and air molecules to move freely around the room. Give it a minute for the molecules to mingle and then get them to freeze on the spot. Ask them to discuss with the student next to them what the role play is showing. For a less active version, get the students to use a combination of milk and white chocolate buttons to model the particles and how they change.

Make sure you use some kind of plenary that checks for relatively deep understanding of the topic. One idea would be to get students to write 'Fact' on one side of their mini whiteboards and 'Fiction' on the other. You then read out statements for them to agree or disagree with.

Techniques for developing scientific understanding

Once the students are clear about the nature of a particle, it is then time to develop this further into what a particle actually is. Atoms, elements and molecules must now be introduced.

The stories of Rutherford and the European Organization for Nuclear Research at CERN are a good place to start. Kick off with how Rutherford set out to test the plum pudding model with his scattering experiments to come up with our current understanding of electrons, neutrons and protons, and then move on to how current work at CERN is trying to understand this model even further (don't hide from the real complexities but don't expect them to understand them all either!). Brian Cox's short film, *A Crash Course in Particle Physics*,[15] will do this for you if you wish: it includes lots of information, so think about your class and how much of the film they can watch before they become overloaded with ideas.

At this point, establishing the *size* of an atom is essential. Get the students to research into different models and then present their findings to the class – this will ensure that you don't move off this concept too quickly. Introduce the basic accepted model of the atom (but do hint at further complexities). The students can then make physical models of atoms to embed the learning.

---

15   See https://www.youtube.com/watch?v=HVxBdMxgVX0.

Techniques for developing mastery

Here are some ideas for discussion or project work:

- If atoms are empty space, why do you not fall to the floor when you sit on a chair?

- Why is ice lighter than water?

- Is solid, liquid and gas all there is (i.e. the importance of plasma)?

You can also introduce what is happening at CERN: the search for Higgs boson, black matter, Bose–Einstein condensates, quarks and leptons. There is no ceiling to what you can teach here, and it can really cement earlier ideas.

An understanding of particles will underpin much of the further study of chemistry and physics, so your curriculum will need to identify all the other points when this topic should be referred back to in your teaching.

ENERGY

Energy is a classic example of where students' everyday use of the word, and their consequent understanding of its meaning, makes this a challenging subject to teach. Students often have the simplistic misconception that energy is something you get when you eat or burn fuels, so when you introduce the idea that energy cannot be created or destroyed the students can begin to struggle. Energy is a mathematical concept and students aren't really up to the maths in this area until A level.

However, energy comes up in pretty much every topic of biology, chemistry and physics (e.g. food webs, photosynthesis, respiration, rates of reaction), so your students will need a 'good enough' model of it. Energy transfer is one of the most important concepts in physics as any change in the universe requires an energy change of some sort, whether this is through photosynthesis, light or electricity. The use of a transference model of energy, as opposed to there being different types of energy, is a useful way of exploring energy up to Key Stage 4.

Foundation activities – ideas to get started with

At Key Stage 2, children are taught that energy is provided by food, that there are various types of energy in day-to-day life, and that renewable energies exist and we should all try to save energy. There is plenty of scope for lots of misconceptions!

The classic start with younger students at Key Stage 3 is an energy transformation carousel practical to elicit ideas about the different types of energy (although 'fuel' might be a better description). Create around 15 different stations around the room and put the students into pairs. At each station have an object or two and get the students to identify what energy types are involved (e.g. hairdryer, bat and ball, radio).

Follow this with tasks regarding 'lost' energy. Stress the fact that energy is not lost but it becomes so dispersed that it can no longer be useful. I would urge you to read Robin Millar's 'Teaching About Energy' paper in which he outlines why you need to discuss energy as a 'quasi-material substance' which can be transferred from one object to another – it is simply unavoidable to do otherwise.[16]

---

16 R. Millar, *Teaching About Energy*. Department of Educational Studies Research Paper 2005/11 (York: University of York Department of Educational Studies, 2005).

At this stage, the aim should be to educate the students into being members of society with an understanding of some of the implications of energy rather than confusing them with the bigger ideas. Your teaching should include energy 'stores' (such as fossil fuels), the impact of global warming and the importance of energy efficiency. In addition, energy should be linked across the curriculum with ideas about food chains and webs, photosynthesis and energy in reactions. Plan your lessons around this initially, as other ideas can be visited later.

Activities might include:

- Compare the energy information on food labels.

- Do an investigation into the caloric value of different foods. This is also a really good way to teach the difference between reliability and validity. Different groups of students can get concordant results in these types of experiments, therefore giving the appearance that the results must be 'correct'. But in reality, because so much energy is dispersed to the environment, the results are unlikely to be valid.[17]

- Renewable energy resources and their advantages and disadvantages.

Techniques for taking it further

The next step is to develop the idea of energy in a mathematical sense. For the most able students there is no reason why this should not happen in Year 7.

---

17 For an experiment on the energy values of food developed by the Nuffield Foundation and the Royal Society of Chemistry see: http://www.rsc.org/learn-chemistry/resource/res00000397/energy-values-of-food?cmpid=CMP00005022.

The idea of 'work done' needs to be introduced first so that the joule can be understood – this is because one of the easiest ways to define energy is as 'the capacity to do work'. From here, various calculations for working out, for example, kinetic and potential energy can be calculated. Again, the key idea that should be taught here is the conservation of energy because the maths can be used to support the concept.

Even though this requires the manipulation of mathematical ideas, you still need to make sure your lessons bring energy to life. Some practical ideas include:

- Use the school sports hall or running track to calculate the kinetic energy each student exerts when running.

- Calculate the 'work done' and the power of students running up a flight of stairs. Turning this into a 'Who is the most powerful?' competition makes students really want to complete the maths work!

- Calculate the gravitational potential energy of a student standing on a chair and then work out the kinetic energy they have when they jump off. Discuss in a think–pair–share activity why not all of the potential energy would have been converted into kinetic energy.

- Work out how much chemical energy was needed for these activities and calculate how many sweets the students have burnt off.

Techniques for developing scientific understanding

Energy transforming from one type to another is often found in GCSE exams and is often answered very poorly by students. Whilst the idea that 'energy is never made or destroyed' is introduced from Year 7, frequently Year 11 students will not be able to apply this idea. Over the space of a student's secondary experience aim to ensure that there is a wide and varied look at more complicated examples of energy transformation.

Here are some ideas that you could integrate into your long term plans:

- How the planets were formed from energy released from hydrogen fusion left over from the Big Bang.

- Look at nuclear bombs and how the energy stored in radioactive uranium is transformed into thermal and kinetic energy.

- Study the numerous energy transformations in a power station and look at how new power stations aim to become more energy efficient through reducing energy dispersion.

- Make your examples relevant to your class – for example, look into energy transformation in a mobile phone (focus on how hot they can get, especially the metal backs of iPhones) and why this affects battery life.

Students can sometimes understand the idea of energy transformation better when maths is used, so plan to:

- Introduce Sankey diagrams for the transformations you have studied and get the students to explain what they mean to one another (to ensure that they understand the implications of the numbers rather than just completing some algebra).

- In chemistry, teach chemical energy in terms of chemical bonds so the students can understand how fuels and food have energy.

- In biology, spend quality time on pyramids of biomass and number as the energy 'loss' here is often taught as an aside.

Techniques for developing mastery

As mentioned above, truly and properly understanding energy at Key Stages 3 and 4 is not really possible due to the complex maths required. So, whilst it may be possible to define energy at this point, it is practically impossible for the students to fully grasp the topic, let alone reach any version of mastery. Instead, for those students who are ready to have their eyes widened, my advice would be to capture their interest in studying physics further.

Here are some ideas you could try:

- A great starting point is the book *How to Teach Quantum Physics to Your Dog* by Chad Orzel which includes some fascinating examples with which to engage students.[18]

- Look at quantum tunnelling (the phenomena where a particle tunnels through an energy barrier it could not classically surmount) and consider the impact this could have for nuclear fusion.

- Explore energy resources in much more detail, covering topics such as fracking and its opponents and the politics of energy.

- Introduce string theory and the reasons some scientists oppose it.

- Go further into detail about what actually happened in the Big Bang.

---

18   C. Orzel, *How to Teach Quantum Physics to Your Dog* (London: Oneworld, 2010).

CHARACTERISTIC 2: SPACING IT OUT

Once you have identified the main hinge concepts for teaching science, the next step is to develop a medium term plan that helps the students to build connections between the areas and develop the skills and knowledge they need. One of the techniques that makes the most difference to student understanding is interleaving topics – leaving gaps between teaching parts of a topic and spreading the learning of each area over time.

Our understanding of this strategy comes from the Bjork Learning and Forgetting Lab in Los Angeles where, for some decades, researchers have been looking into how memories are formed – their findings are essential reading for teachers.[19] In one study, participants were asked to learn the formulae for calculating the volumes of various solids. Those who were taught these in a blocked massed practice (i.e. all the subject in a block before moving on to other subjects) remembered only 20% of what they were taught. Those who were taught for the same amount of time, but with that time spaced out in smaller chunks over time, remembered 63%.[20] A clever science teacher will use this spacing effect in curriculum planning.

What the Bjork Lab research shows is that if knowledge and skills are used repeatedly over time, spaced out with gaps, then it is more likely they will be retained than if they are taught in a block without breaks. This makes sense for any of us who have ever crammed for an exam: you might remember information for the exam itself but quickly forget it afterwards. Thus the practice, practice, practice mantra does not always make complete sense and is, at the least, under-nuanced. What this means in terms of your planning is that if you teach evolution for a term, but then do not return to it for over a year, the students will have forgotten much of it, regardless

---

19 See http://bjorklab.psych.ucla.edu/research.html.
20 D. Rohrer and K. Taylor, 'The Shuffling of Mathematics Practice Problems Boosts Learning', *Instructional Science* 35 (2007): 481–498.

of how much mastery they achieved when it was first encountered. What works rather better is spaced repetition which enables us to control students' exposure to the topic over time. Mastery cannot be retained through one chunk of teaching!

This idea goes against the spiral curriculum that schools have been using in recent times, where short topics are taught as blocks once a year and each year the topics get slightly harder. When most students can't remember the difference between respiration and breathing by the time they get to Year 11, then there is good reason to reflect on the way we structure the curriculum.

An example of a traditional Year 7 and 8 science curriculum looks like this:

Year 7	Year 8
Cells	Food and digestion
Acids and alkalis	Atoms
Energy	Heating and cooling
Reproduction	Respiration
Chemical reactions	Compounds
Electricity	Magnets
Environment	Microbes
Particles	Rocks
Forces	Light

Year 7	Year 8
Variation	Ecology
Solutions	Rock cycle
The solar system	Sound

This is based on the old national curriculum, where four topics were studied each year for biology, chemistry and physics. This material was structured in a spiral so that, in theory, topics were introduced in Year 7, then a related topic was covered in Year 8 and so on. Each topic was taught and assessed separately. Thinking back to the idea of hinge concepts, it is clear that there are too many topics in this example for the students to really grasp and master the basics of science, resulting in a dizzying rush through each topic with little or no time to draw links between them.

It's actually relatively easy to take the ideas of threshold concepts and spacing into account when planning a new curriculum (as shown in the example below). And the benefit of teaching in this way is that the three sciences are taught much more seamlessly by focusing on the key hinges that underpin them. This removes one of the main criticisms that students often have about science in comparison with skill based subjects such as English and maths: they complain that what they learn is just a disparate list of unrelated facts.

	Autumn term		Spring term		Summer term	
Year 7	The basics of life: cells, variation and evolution	The build-ing blocks: particles	Energy	Humans: organ systems and how they are adapted to their function	Atoms and the periodic table	Forces and motion
Opportunities to revisit	Use a pre-assessment (see Chapter 8) to under-stand where to start. You may be surprised at how much some students have learnt in primary school.	Ensure you deal with any misconcep-tions (these are identified earlier in this chapter).	The impact of energy resources on global warm-ing and the implication for evolution.	The evolution of these struc-tures including comparisons (e.g. the pen-tadactyl limb in various mammals).	Moving forward from particles, recap thoroughly before starting on atoms.	How the Earth stays in orbit. Forces in the body such as the heart. The link between forces and energy.

	Autumn term		Spring term		Summer term	
Year 8	Interdependence: the environment, including parasitic and mutualistic relationships	Bonding	Electricity	Genetics and DNA	The Earth	The solar system
Opportunities to revisit	Evolution: the impact of environment on speciation. How and why these relationships exist. The impact of fossil fuels on the environment.	Start with a complete refresh of the periodic table and misconceptions around basic particle theory. Re-emphasise key words throughout.	Why our need for electricity is causing issues with global warming. The importance of efficiency not just for electrical items but for animals as well.	The impact of the environment on our DNA. How DNA and genetics result in evolution.	Links with the main compounds on Earth, resulting in rocks and why they form as they do. How the rock cycle affects the environment.	Forces and energy. The Big Bang and how this has affected the Earth and the compounds present.

A good curriculum must identify ways in which we can link important ideas together within an overarching story. This does not mean simply mentioning that story in passing but taking time to revisit it and look into subjects further and more deeply. For example, when teaching about efficiency in physics, it could be that a lesson is spent looking at the efficiency of penguins huddling together in groups and why, therefore, evolution has resulted in them living in cooperative groups. Making such a link does not qualify as simple repetition, which will only get the students so far. Instead, it gives them a new way to think about the same topic, which is what you need to do to develop those elusive thinking skills. This may make the pace of progress feel slower at the time, but the long term gains in memory retention are worth it.

One of the other ways to ensure spacing and revisiting happens is to build it in to how students are assessed over time, otherwise you can't be sure that it is working. (The ideas behind this are covered in more detail in Chapter 8.) A good curriculum should also identify when all the different skills required for science are taught. This is of particular importance as science teachers have a tendency to 'front end' their skills teaching: in other words, they focus on the skills of planning an experiment rather than the skills of analysis and evaluation.

None of this is particularly novel. Some of the best teachers I know do not constantly teach 'wow' lessons, but they are successful because they use these methods intuitively to ensure that their students retain information. By thinking about these practices before we start teaching in September, we can make sure they are properly well thought out.

CHARACTERISTIC 3: MEETING THE NEEDS OF ALL

The curriculum is, to a certain extent, merely an idealised way of teaching. It is clear, however, that it should also allow for the range of differing individuals we teach to flourish. The way in which you modify the curriculum for your classes and for the individuals within those classes will impact on how successful your teaching is.

The dos and don'ts of differentiation

Do have the highest expectations. Every student needs to be challenged and struggle is necessary for them to learn. Ensure that students know what excellence looks like for each task.

Do know your students well. Mark regularly and check for understanding as you go.

Do set up learning routines so the students know what to do when they are challenged by a task. What do the students think they should do when they are stuck? Do the students know how to improve? How do you insist on students working to the best of their ability? Are task ladders used? Or bronze, silver and gold tasks?

Do use multiple models to teach difficult concepts, so if the students don't understand one mode of explanation then there is another one easily available.

Do use scaffolding, but aim to remove it as they progress.

Do think about the literacy content of your lessons. Know your students' reading ages and set them appropriate levels of challenge.

Do use feedback wisely. It is the easiest way for you to know your students and for them to know how to improve.

Do use a seating plan to your advantage. Think about whether it is best to use peer or mixed ability groups. Consider having two seating plans that you can switch between for different types of activity.

Do use data, but be wary of it. Data can help in planning – for example, a pre-assessment can support differentiation for the whole class by allowing you to know the ins and outs of your students' understanding. Equally, data can mask things that you do not know about and lead to a lowering of expectations for certain students.

Don't feel that differentiation needs to mean hours spent planning different resources for each ability level for every task. Differentiation can be as simple as modifying the way you interact with different students.

Don't rely too much on extension work. Some students will just see it as 'extra work' and will neglect to do it. Instead, think about making tasks more open ended and provide thinking questions to extend their ideas. These are questions that push the students to think about previous learning and apply it to new situations – for example: how might this be used in another situation? How could that problem be overcome? What patterns could lead to an alternative answer? Why do you think this works? Do you think it will always work? How might you argue against your answer? What are some of the complexities that we should consider? What else would you need to find out to make this information more consistent?

Science can be a difficult subject to differentiate for, as even the most able students need to learn some basic ideas at the beginning of each topic so they can move on to doing something with what they have learnt. But this does not mean that it is impossible to meet the needs of the students in your classroom. Many of the ideas in Chapter 4 explain how to differentiate for different ability students with very little effort, so provided you use a variety of tasks and think about how to challenge each student, you will manage this.

Chapter 6
TEACHING DIFFICULT TOPICS

'I CAN'T DO SCIENCE'

This chapter will look into those topics that, let's admit it, some of us don't like teaching very much. These topics include electricity, heat, moles, osmosis and maths that can be a bit boring or difficult to teach. What most of these areas have in common is that they are rather abstract, they require a quite complex bringing together of different ideas and the students tend to have only very limited previous knowledge of them. And, sometimes, it's not just the students who are confused by these subjects – it's us! No matter how tempting, the solution is not to ignore them but, instead, to teach them especially well and in great depth. The starting point with the difficult topics is to really emphasise to students why they are important so they can link them to everyday life.

ELECTRICITY

Research by the University of Oxford has found that students who were given a mild electric shock were better at completing maths puzzles than those who'd had a placebo.[1] It's perhaps something of a shame that this method is deemed unethical in schools.

---

1   A. Snowball, I. Tachtsidis, T. Popescu, J. Thompson, M. Delazer, L. Zamarian, T. Zhu and R. C. Kadosh, 'Long-Term Enhancement of Brain Function and Cognition Using Cognitive Training and Brain Stimulation', *Current Biology* 23(11) (2013): 987–992.

When I first started teaching, electricity was the topic I most disliked (a quick survey of my colleagues confirmed the same). I just didn't enjoy it as electricity is such an abstract idea and quite simply something which didn't interest me. (However, I've recently bought a new vacuum cleaner and had a fascinating educational conversation about voltage and wattage with my husband, so understanding electricity turns out to be useful after all!) After teaching it year in, year out – and learning from my errors – that feeling has (mostly) gone away. One of the chief issues is that, not only do students have misconceptions about what electricity is, so do many textbooks which often confuse electrical energy and electricity (and we've already looked into how confusing energy can be in Chapter 5).

I teach this topic by rooting it in real life applications, not just for the sake of the students but also for myself. Whilst electricity might seem easy conceptually (after all, how long would you cope in a blackout?), it is easy to get bogged down in the scientific understanding. One way to do this is to teach the whole electricity topic under the guise of 'How does my mobile phone work?', as mobiles and food seem to be two of the three things that teenagers think about the most. However, the problem with most schemes of work is that they teach electricity the wrong way round – they jump straight into circuits and symbols and confuse the students by introducing all the key words at the same time. Instead, to ensure full understanding, the best way I have found to teach electricity is the way history tells it.

To start with, what do these three have in common?

These three images tell the story of how electricity – one of the most fundamental aspects of our life – was discovered. As a class, we discuss how static electricity and lightning has been known about (but not understood) for a long time and how Galvani discovered 'animal electricity' (as he called it) in frogs' legs. Before moving on to circuits, we focus on the nature of electrical charge and magnetic fields. When we have covered electrical charge, and I'm sure the idea of electrons repelling each other is fully understood, we then look at how electrons behave once the charge is separated. The next step is to look at Faraday's experiments which demonstrate how electricity is generated, as we essentially generate electricity today in the same way as Faraday. A nice practical for students involves using a metal coat hanger, a 6V battery and a compass to demonstrate that an electric current produces a magnetic field.

When you have covered this material, the more typical scheme of teaching can then be followed.

A good role play!

Analogies and role play are really useful ways to teach the abstract idea of electricity. Use coloured string to represent copper and create a loop of this around the sports hall or your classroom. You'll need two students to hold either end of the string – they should also hold hands to represent the positive and negative parts of the battery. The remaining students are electrons and should wear negative signs on their tops to remind the class of the reason for their movement. They should move towards the positive side of the battery.

Once you have set up the basic idea for the circuit, it can then be used to demonstrate most areas of electricity. You can use the string to make parallel and series circuits and model the differences between current and voltage. You can remove the battery and put in an alternative current, forcing the electrons to keep moving in different directions. You can demonstrate conservation at a junction by using student 'ammeters' (students who make a beep each time an 'electron' passes by). Finally, you can add a resistance tunnel by using students to get in the way of the movement of the electrons.

Linking back to the idea of teaching this unit around the mobile phone, you can discuss the battery and how alternating current is converted when it is charged. You can also look at the inside of a phone, getting students to make predictions about whether the circuit is in parallel or series and discuss why phones get hot (and burn your ear) when you use them.

Practical work with circuits

In practice, getting students to understand ideas about current and voltage can be quite difficult, even if they can recognise electric symbols from their primary school education and can draw simple circuit diagrams. The issue most science teachers face is how to get students to actually make a circuit so they can observe some of the ideas you need to teach them. Frequently, for some strange reason, students are unable to transfer their knowledge of a circuit diagram to a practical set-up in the lab.

You might remember the first time that you taught circuits. I certainly remember mine: a group of boys made what they called 'super electricity' by creating a circuit with as many batteries as they could find. Most other groups were insistent that they had made a parallel circuit but, in fact, had just made a very complex series circuit by using at least twice as many wires as they should have. Since then I have taught circuits slightly differently. On students' first exposure, I ask them to draw out the circuit they want to make on a sheet of A4 and then get them to assemble their circuit on top of it, starting with the components and completing it with the wires. Once the students have mastered this method, the diagram scaffold can gradually be withdrawn.

An idea to reinforce this learning is a 'circuit race'. Display different circuits on the whiteboard and ask the students to race against the clock to make them. Award three points if they manage it in 10 seconds, two points for 15 seconds and one point for 20 seconds.

Some of the ideas within electricity need to be taught in as many ways as possible and most really benefit from practical work. Take your time over these big concepts otherwise you might end up with students who have just a superficial understanding of what you've taught them.

HEAT

Heat is one of those topics that, at first, seems as if it's not that difficult. There are nice practical experiments to demonstrate the big ideas and it's a topic that is really easy to link with everyday life for students. Think again! Students find the topic of heating really tough. If you don't check for understanding on a regular basis, you will ask them to complete a test at the end of a topic and get answers such as 'heat rises', 'particles expand', 'cold can move' and, my all time favourite, 'heat particles'. The main way to stop this is to pre-empt problems and plan to avoid them.

In the previous chapter, we looked at the idea of particles being a key lever for student understanding, and that's definitely the case here too. Do not even think about teaching heating without first going right back to particles and how they behave. At this point, if you haven't done so already, it's important that you move away from the idea of the general 'particle' and move towards atoms and molecules, because questions will refer to specific examples and, to avoid later confusion, you should make sure it is taught explicitly now.

There are various ways to teach students about heat, but first you have to get rid of those nasty misconceptions. Start with something like this: find around 10 different images of moments from everyday life that involve heat transfer and put these up around the room. Get the students into groups and then carousel them around the pictures. At each image they need to think about and discuss the same questions:

• What is happening in the image?

• Where does the heat come from? Where does it go?

• What questions do you have about the image?

Be clever with the photos by choosing ones that will challenge misconceptions – for example, someone who is cold in a draughty room, illustrations of particles changing state, an image to illustrate the idea of metals being cold. You can also use concept cartoons here (see Chapter 2).

Use the start of the topic as an opportunity to go over the difference between heat and temperature. Try to use imaginative titles for these lessons such as:

• Does a blanket have its own warmth?

• How do astronauts cope with the cold?

• How can people walk on hot coals?

Capturing students' attention is important because they will often believe they already understand subjects like heat because it's part of real life, and so do not always engage with it.

Once you have recapped particles, energy and dealt with any misconceptions, you can teach conduction, convection and radiation. Don't fret if you've had to use up a good few lessons before you get there. This is another area that really benefits from 'over teaching' the basic concepts. Rather than just showing students some standard practicals (e.g. fitting a metal ball through a hoop, black and white cans), do as many different ones as you can come up with.[2] Pay heed to Chapter 7 on practicals, though – there is no point in doing this if you don't go over the science behind each experiment.

A great way to recap conduction, convection and radiation is to head to the food technology room and grab some pans, popcorn, a popcorn maker and a microwave. Put the students into groups and give each group one way of making popcorn (on the stove, in the microwave or in a popcorn maker). Then get them to explain how their method works using the scientific principles you have taught them when they are feeding back to the class (perhaps using role play). As an aside, you can also use popcorn to illustrate radioactive decay which is a really good way of showing that this is a random process – and you can let them eat it at the end!

MOLES

Another topic that is abstract, not directly observable and requires some maths teaching is the mole. This is typically taught only towards the end of Key Stage 4, but that doesn't stop it being one of the most difficult topics to

---

2   For example, make paper 'snakes' to demonstrate convection (http://www.science-sparks.com/2011/09/12/convection-snakes/), blow up plastic balloons to show conduction (http://science.wonderhowto.com/how-to/perform-cool-water-heat-conduction-experiment-419926/) or try out the brilliant examples for using a microwave to demonstrate radiation at: http://www.scienceinschool.org/2009/issue12/microwaves. For more thermal transfer experiments visit: http://practicalphysics.org/thermal-transfers.html.

teach. There are four areas to address well before you even consider introducing any mathematical equations:

1   Check that their previous understanding is solid – at this point, it is essential that students fully grasp particles, chemical formulae and balancing equations.

2   Macroscopic first. To be able, eventually, to complete all of the maths required, a conceptual understanding of the mole is vital. The best way to make this real is to start with what the students can see so they get a feel for the mole. Start with why Avogadro's number is a necessity – that is, because we need something to measure that which we cannot see. Use some day-to-day examples (e.g. a ream of A4 paper, a dozen eggs, a kilogramme of flour) and use props (e.g. a jar of sugar and four reams of A4 paper). Ask the students how much paper and sugar there is; they will respond in terms of grammes, reams, etc. Discuss with the students why we need to have these measurements and get them to come up with further examples. Use the historical context of Avogadro – what he was trying to work out and how no one believed him because they couldn't get their head around what he was claiming (mainly because it was theoretical and some data at the time refuted his findings).[3]

3   Use practicals. There are many hands-on ways to make the topic come alive. Get the students to prepare one mole of water, salt, iron filings, etc. and keep these on display throughout the topic. Or go to the food technology room and get students to make fruit salad with equal 'moles' of cherries, blueberries, etc.

---

3   For a concise description of Avogadro's work visit: http://www.rsc.org/chemistryworld/issues/2006/march/avogadro.asp.

4   Make sure they understand the scale of the number. Use analogies so that 6.02 x 1023 means something to them (e.g. one mole of tennis balls would equal the size of the Earth).

Take your time on this as it can be easy for students to find themselves in information overload. Once you have taught the mole sufficiently well, and in enough detail for the concept to be understood, then you can move on to the mathematical equations. Start with simple one-step questions before you progress to questions that require problem solving. To be successful here, you will need to really spell out to the students how you go about it, emphasising and modelling the metacognitive processes required. One of the key issues in problem solving-style questions is that it's hard for students to see past the sheer amount of numbers they need. By teaching them how to organise the information clearly you'll find that they grasp it quicker.

Here is an example using lead nitrate:

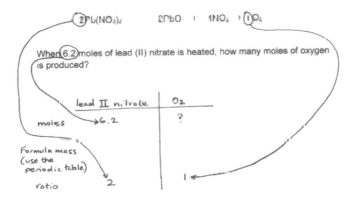

Also, you might consider not teaching molar mass, volume and concentration and gas volumes all at the same time. Instead, space them out over the term so the students don't get mixed up between the three ideas.

OSMOSIS

Osmosis is yet another topic that has an understanding of particles at its heart. Begin by making sure that your class are up to speed with particles (particularly if you are teaching biology as a separate subject rather than combined sciences).

Osmosis is a subject that students seem to be okay with initially, and the first time you teach this you may reflect that it has gone fairly well. A traditional textbook way of teaching osmosis would be a teacher explanation using diagrams to illustrate what is going on at the particle level, followed by a class practical to examine the effect of different concentrations of salt water on a mass of potatoes. At the end of these two lessons, it is likely that the class will be able to regurgitate the definition of osmosis and explain their own experimental results.

But give the students some data on a similar but slightly different example (e.g. cucumbers as opposed to potatoes) and wait for hands to pop up all around the room with students confused about whether the cucumbers would gain or lose mass. This is an example of superficial learning. Basic assessment for learning techniques can really let you down when it comes to identifying if students have fully grasped the topic. To eradicate this problem, you need to give this topic sufficient time so the students can understand the links between all of the different examples of osmosis you're able to come up with. Make sure these cases are rooted in everyday life.

Some questions that you can get the students to answer during this topic include:

- Is drinking lots of water good advice? What happens if you drink too much?

- Why do sports drinks always contain electrolytes? Do athletes really need these?

- If you are stranded on a desert island, why can you not drink seawater?

- Why does a plant become limp when you haven't watered it?

- Why does eating too many sweets or sugar-free gum give you the runs?

- How does kidney dialysis work?

Then move on to a reminder of diffusion. It is likely that osmosis is being taught in relation to diffusion and active transport in plants, but I've frequently seen students experience cognitive overload from being taught all three processes at the same time. I find it better to teach them separately and then bring them together once the students fully understand each concept individually.

What we need at this stage is for the students to observe osmosis and explain it in a variety of ways. Use simple demonstrations but avoid anything to do with plants. Here are some suggestions.

What's happened to those eggs?

I like to start this experiment before going into lots of particle diagrams (which can be very dull). You will need some hard-boiled eggs, which the students peel and soak in vinegar for at least half an hour. Whilst they are doing this, ask them to think about what the vinegar is doing by bringing ideas from chemistry into play.[4] Then get them to remove the eggs from the vinegar and weigh them. Leave the eggs overnight – one in a jar of water and one in a jar of sugared water. In the next lesson, remove and dry the eggs and then weigh them. Use a think–pair–share to encourage them to discuss what has happened.[5] You can then move on to the more standard explanation of osmosis (be careful to plan this out) and the standard potato in saltwater experiment.

Next, get the students to do a similar experiment completely independently – this is the ever controversial 'enquiry-based learning' in action. But this time they must plan an experiment into how salt or sugar levels affect the size of gummy bears. Once they have completed the experiment, they should write up their findings in detail and unaided. Use this as a mastery assignment: insist that the students redraft their work and do not mark it until it is near perfect.

Finally, if you have been teaching osmosis in relation to active transport and diffusion, make this really clear. A Venn diagram for students works really well here.

---

4   As it's an acid, the vinegar is removing the final bits of the shell to allow osmosis to happen.
5   An example of this experiment can be found at: http://www.nuffieldfoundation.org/practical-biology/investigating-osmosis-chickens%E2%80%99-eggs.

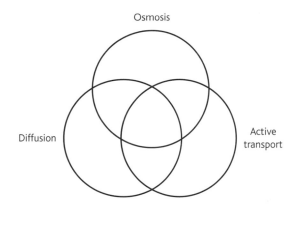

HOW TO TEACH EMOTIONALLY CHARGED TOPICS – A GUIDE TO REPRODUCTION AND THE LIKE

Teachers have a duty to ensure that our students are not just well versed in academic subjects but also in life skills. Depending on your whole school policy, it is likely that you will be expected to teach the following challenging areas:

- Puberty

- Reproduction, sex organs and female genital mutilation

- Contraception, including how the pill works

- Sexually transmitted diseases

- Drugs (legal and illegal)

- Healthy living, including obesity, anorexia and bulimia

Given the profoundly important consequences that lack of knowledge in these areas can have on the success (or not) of your students' lives, it is simply not good enough not to teach these areas well or not to address both the teacher's and the students' potential embarrassment. Too many young people leave school believing in myths that can have catastrophic effects on their lives (a 2011 survey by the sexual health charity Brook revealed that 33% of 14–18 year olds believe that you can't get pregnant the first time you have sex![6]), so we have a duty to make sure they know *all* the scientific facts.

Here is some advice on dealing with emotionally loaded and tricky subjects:

- Remember that you are teaching this in a science context. This means that technical language is required and should be modelled right from the start.

- Plan your ground rules for discussions and make them crystal clear to the students. Work on these with your class and come up with some 'dead in the water' statements that can't be said from the word go. One of these needs to be about not asking personal questions of individuals. (If not, they are going to ask them of you, and it will be difficult to balance being open whilst maintaining appropriate boundaries.)

- When planning, consider the ethnic, religious and sexual diversity of your group. How will this affect how you set up the discussion? Think also about how you would deal with any discriminatory opinions

---

6   See E. Saner, 'How Good is Sex Education in Schools?', *The Guardian* (10 October 2011). Available at: http://www.theguardian.com/lifeandstyle/2011/oct/10/how-good-is-sex-education.

students might have. Hopefully your school will have a clear policy on dealing with these issues and your school culture will mean that such statements are unlikely to occur, but they could. Never ignore prejudicial statements as this could be seen as giving your approval. Challenge the student, being careful not to humiliate them, employing information and facts to challenge simplistic views. At times, a simple statement in class followed by a private conversation later on may work best. If you are worried, discuss this with your head of year or department.

- Give straight answers with a straight face. Your students will follow suit.

- Treat sex as a normal healthy part of life, but touch on the importance of consent and relationships. (This is likely to be taught in more detail in PSHE lessons so your role will most likely be to focus on the biological sides of these topics.)

- Consider starting to teach sexual disease or reproduction with a look at reproductive organs! This can help the students to get used to you using words such as 'penis' and 'labia' before you need to broach anything more sensitive, by which time you should have created the right kind of atmosphere.

- Ensure the students know that if they would like to speak to someone about any of the issues that come up in your lesson then they know who to go to (i.e. your nurse or school counsellor) and they should feel comfortable in doing so.

The question section in Chapter 2 provides a number of techniques for holding purposeful discussions. Here are two more that are specifically useful for difficult discussions.

The question box

With these types of subject, you can be inundated with questions from *very* interested students, some of which it would be sensible for you to think about how you will answer in advance. Invite the students to write a question on any of the topics above and drop it into your question box at the end of a lesson. Answer them in the next lesson.

Opinion finder

When covering a topic that students have a number of different ideas about, try using an opinion finder. Set a question such as, 'Should abortion be allowed?' Ensure you have given enough background information into the intricacies of the issue first so they are armed with the scientific, emotional and psychological evidence they need that might affect their opinion. Then ask the students to discuss the statement with their peers and record their results in an opinion finder like the one below. For your plenary, get the students to discuss what they have learnt from one another and then get them to reflect if their opinion has changed at the end.

My opinion at the beginning of the lesson ...

	Tally	Frequency	Reasons
Strongly agree			
Agree			
Disagree			
Strongly disagree			

My opinion at the end of the lesson ...

MATHS

Teaching science is tough. Not only do we have to deal with the vast array of health and safety issues in our labs, cover a huge amount of content and teach literacy skills, but we also have to be maths teachers. We have to embrace this as maths is itself the science of patterns and because maths is a fundamental part of any scientist's daily life. It is impossible to separate science and maths, so a greater emphasis on maths in the classroom is advantageous not only to further student understanding but also as a way to further embed their mathematical and analytical abilities.

STEM subjects (science, technology, engineering and maths) are increasingly important in the global economy, and one of the ways in which the government is trying to improve this is by making maths more challenging for students. This doesn't just affect maths teachers, however, as one of the tests of whether a student has a good enough understanding of numeracy is their ability to apply it to new situations. This is where science teachers become saviours. In the GCSE specifications for first teaching in September 2016, the maths content in exams has been expanded so that it will now constitute 10% of the biology paper, 20% of the chemistry paper and 30% of the physics paper,[7] thus making it even more important that we explicitly teach maths in our classrooms.

Many students tend not to love maths, so when they walk into your classroom expecting a practical, some interesting facts and a fun vocabulary game, only to find they are presented with maths problems, they can be a little disappointed – that is, unless you have created the right culture in your class. The Association for Science Education (ASE) raised concerns recently about how well the mathematical aspects of science are understood by

---

7   See http://www.aqa.org.uk/about-us/what-we-do/policy/gcse-and-a-level-changes/
     changes-in-your-subject/science-changes.

students,[8] so this is an area that is becoming increasingly important for science teachers to be well versed in – specifically in how to teach it.

Is your maths up to it?

The standards expected for a student finishing Year 11 have probably risen since you were at school, so it is worth checking if your maths skills are up to the basic level expected of a 16-year-old. Try this problem which your average Key Stage 4 student should be able to answer quite easily:

This is Kim's garden. She needs to cover the garden with compost, which costs £1.50 per square metre. The garden includes a square pond that is 5.5 metres wide. How much will it cost in total to put compost in her garden? Give your answer to two decimal places.

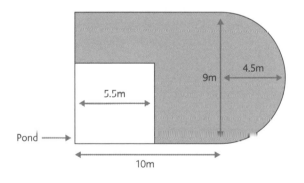

---

8   R. Boohan, 'The Language of Mathematics in Science', *School Science Review* 360 (March 2016): 15–20. Available at: https://www.ase.org.uk/journals/school-science-review/2016/03/360/.

a   £137.31

b   £325.85

c   £132.03

d   £132.04

If that is a struggle for you, or if you think there are other areas you may need to work on, then spend some time with the very nice members of your maths department as it is vital that your skills are up to the expected standard. (The answer to the question is at the end of the book on page 207.)

New maths methods

There's no shame in struggling with maths – even the great physicist Ernest Rutherford, who established the nuclear structure of the atom, wasn't that good at it. In lectures he would often get lost in his own equations and need a break, or he would hide his confusion by asking his students to work it out for themselves (I do not advise this as a teaching method!). However, finding maths hard, and consequently not teaching it, is simply not an option in the modern science classroom. It is essential that we up-skill our mathematical content knowledge, and also that we teach it in a way that is consistent with the way maths is now taught. Depending on your age, it is very likely that the way you were taught maths is very different to how it is taught now.

The following is a quick list of the new (and some not so new) methods that are used to teach students basic maths skills from primary schools upwards. If you're unsure about any of them, it is worth getting the maths department to update you on how they work and what the rationale is behind their use.

Number lines

This is a mental method to support students with addition, subtraction and negative numbers. Like most good maths techniques, it provides students with a structure for decomposing and recomposing numbers. You were probably taught how to use vertical layouts when you were at school but many students make errors with these. The use of number lines provides them with a way of thinking about the maths that doesn't necessarily require a pen and paper, and it gives them stepping stones to complete the calculation.

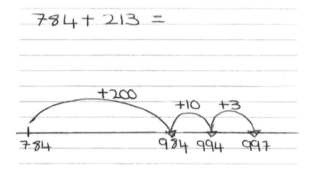

A number line allows students to visualise the addition taking place. To do this, sketch a line and jot down the first number on the left. Next, split the number to be added into its constituent parts to make it easier to deal with – in this example, 213 is broken down into 200, 10 and 3. Then use the number line to add the two together: first jump forward from the starting number by 200 to make 984 followed by jumping the 10 to 994 and finally the 3 to reach the answer of 997. This is likely to be how you complete mental arithmetic in your head but some of our students will not be at that point yet, so teaching this method can be really useful to them.

This method is also nice to use in conjunction with compensation methods – that is, finding an easier calculation first.

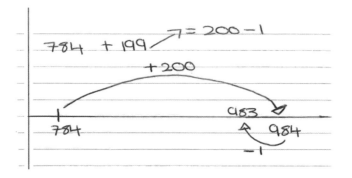

Here, the number line is initially drawn in the same way with the starting number on the left hand side. But instead of breaking down the second number into its constituent parts, it is changed into a number that is easier to work with – in this example 199 is rounded up to 200 which is then added to the initial number. To finish off, the change that was made initially (adding 1 to 199) has to be compensated for, so that 1 is subtracted from 984 to reach the answer of 983.

Grid methods for multiplication

The grid method for multiplication avoids the dreaded long multiplication that many of us were taught in school. Again, it involves breaking down the numbers into manageable amounts in a grid before bringing them back together again.

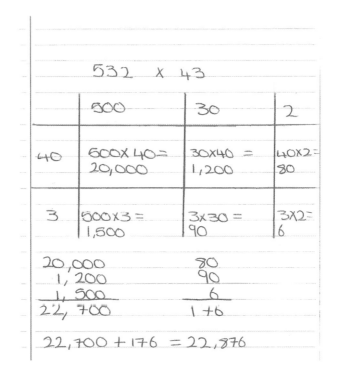

The first step here is to partition the numbers – in this example, 532 into 500, 30 and 2 and 43 into 40 and 3. These are then added to a grid with one number along the top and one at the left. We complete the grid by multiplying the numbers that fall into each box. The answers for each of the multiplications then need to be added together to reach the final answer.

Division

This is a technique for avoiding the long division method that most of us hated as students (sometimes known as the bus stop technique). This method focuses on chunking the numbers to make them more manageable.

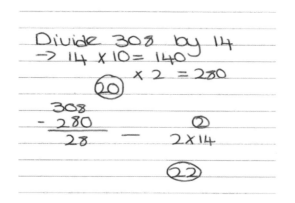

In this example, you know that 14 x 10 = 140 and that two 140s would fit easily into 308. This would leave a remainder of 28, which would be 2 x 14. So you have two lots of 10 from the first calculation and two 14s in the second, bringing you to 22 in total.

Fractions

When teaching fractions, make sure you use the same vocabulary as your maths teachers as this will provide students with a key to access their previous knowledge. The top number in a fraction is the numerator and the bottom is the denominator.

Numerator
_____
Denominator

Percentages

Using ad hoc methods like the following example can be useful:

> In the experiment, 27% of the seeds germinated. There were 250 seeds to begin with. How many have germinated?

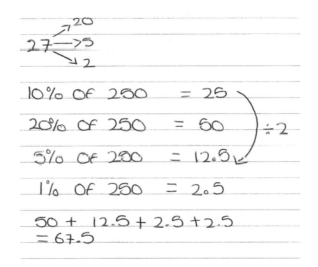

$$27 \rightarrow \begin{cases} 20 \\ 5 \\ 2 \end{cases}$$

10% of 250   = 25 ⎞
20% of 250   = 50 ⎬ ÷2
5% of 250   = 12.5 ✓ ⎠
1% of 250   = 2.5

50 + 12.5 + 2.5 + 2.5
= 67.5

When working with percentages, it is easiest to first work out what 10% and 1% of your quantity is (i.e. by moving the decimal place once or twice respectively) and then use this to work up to your answer. It is easy to determine that 10% of 250 is 25 and therefore that 20% must be 50. The remaining 7% is broken down into 5% and 2% – 5% is half of 10% so is 12.5 and 2% is double 1% and so is 5. Then the amounts are added together to come to 67.5. Always get students to 'common sense check' their answer – e.g. 50% of 250 is 125 and so 25% is 62.5 so it is likely that the answer is correct as it is slightly more than 62.5.

You might be wondering why we bother with these methods given that students can use calculators in their exams. In answer, doing maths using these methods gives the students, particularly the younger ones, more opportunities to use and understand the relationships between numbers. This will help them when they are older as they will be able to estimate and check their calculations. I often use calculators as a checking method once the students have worked things out independently. These methods also build students' confidence in their mathematical ability and support the acquisition of an important basic skill. But the real reason for including these methods here is because it's really important that students are taught consistently and that the science department use the same methods as their colleagues in the maths department.

Maths in science

You should ensure that you are teaching your students the following areas of maths:

- Using (and converting between) decimals, fractions and percentages.

- Standard form – in my experience, this is not taught that frequently by maths teachers so you may need to teach this from scratch.

- Using milli, micro, nano and pico.

- Using ratios.

- Rounding by given significant figures and decimal places, including determining appropriate numbers of significant figures.

- Averages – mean, median, mode and range – and understanding the circumstances in which it is appropriate to use each one.

- Graphs and frequency tables (including histograms).

- Percentiles (which are not often taught in maths classes).

- Sampling methods (again, not taught that often by maths teachers).

- Simple probability.

- Calculating efficiency and rate changes.

- Using inequality symbols ($\leq, <, >, \geq$).

- Changing the subject of a formula.

- Recognising that straight line graphs can be represented by the formula $y = mx + c$ and then understanding how to work out the slope and the intercept.

- Calculating the tangent of a curve.

- Calculating the area, surface area and volume of a cube.

- Inverse square law (for photosynthesis).

Skills wise, you will need to make sure that students can:

- Select and use the right maths techniques to calculate simple problems, including using the correct equations to calculate unknowns.

- Apply mathematics to explain scientific principles.

- Use mathematic skills to solve scientific problems.

Reflect here both on your mathematical understanding and on the pedagogical approaches you need to teach these skills, and also whether this range of topics is reflected in your curriculum design.

Teaching good graphs

In a recent survey completed by the Royal Society of Chemistry, a group of science teachers were given some data and graph paper and asked to plot the yield in a fermentation experiment over time.[9] When the graphs were analysed, the researchers noted how diverse the results were – some teachers swapped axes, some used lines of best fit, others used curves and there was general disagreement on whether the line should go through the origin or not.

If you were to do a similar experiment in your school's science and maths departments, would you find the same thing? I work in a really talented department, but I can recall completing a practice controlled assessment in which we all taught our classes to complete a graph differently. The initial step here is to clarify with your colleagues how graphs should be completed.

---

9   R. Needham, 'The Language of Maths in Science', *Royal Society of Chemistry Education in Chemistry Blog* (13 May 2015). Available at: http://www.rsc.org/blogs/eic/2015/05/language-maths-science-graphs.

Ask yourself, does your curriculum outline when you will teach students to do the following?

- Choose the clearest scale they can. The graph should always fill the page and students should be taught to use a scale that goes up in factors of 2, 5, 10, etc. as it will be easier for them to complete. With some students, you may have to start by giving them graphs with a few hints on the axes.

- Read a challenging scale – for example, a graph that has two different y axes with separate scales (one on the left of the graph and the other on the right) referencing different variables.

- Distinguish between a discontinuous and continuous set of data and how this affects the kind of graph you draw.

- Decide which variable to put on each axis (the independent variable always goes on the x axis).

- How to draw a line/curve of best fit and how to check if it is good (i.e. half of the points above the line and half below).

- When to go through the origin and when not to.

The human graph

Scatter graphs always seem to be difficult for students. One idea here is to try the 'human graph'. Ideally, you will be able to take your class outside onto the field or into the sports hall. Split the class into two, identifying two team leaders. Give each team two balls of different coloured string and a set of data. Borrow the stands that the PE department use when playing rounders and challenge the students to use these objects and their bodies to come up with an accurate representation of their data. Be clear at the beginning on your success criteria and then review it at the end.

Physics equations

The new GCSE specifications include no less than 19 physics equations that students have to be able to recall (plus another eight they will need to be able to use). There are two questions here for teachers: how do you get your students to use them properly? And how do you get the students to remember them?

It is essential to teach the equations explicitly. When teaching kinetic energy, for example, get two students up at the front of the classroom jogging on the spot (choose two students of different mass and fitness and make sure they are the kind of students who can multitask). Ask the class which of the students has the most *kinetic energy* and, through questioning, build up to the idea that it depends on their mass and velocity. Now ask them to run on the spot as fast as they can. Ask the students what is more important, mass or velocity? From here, you can introduce the equation: kinetic energy = ½ x mass x velocity^2. As an aside, when teaching students how to use this equation, refer them back to BIDMAS (which they will know from maths) as it provides them with the necessary literacy needed to use the equation correctly.[10]

---

10   The acronym BIDMAS helps students to remember the order in which to perform operations: brackets, indices, division, multiplication, addition, subtraction (some schools may use BODMAS, with order instead of indices).

Are equation triangles bad?

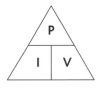

The equation triangle is loved by many a teacher and, indeed, many a student too – but is it a good thing for understanding? They definitely help students to use equations without having to manipulate or remember them. The students simply cover up what they are trying to find out, and if the remaining two variables are side by side then those figures are multiplied, or if one is above then they are divided.

However, equation triangles also work quite nicely to hide students' poor maths skills. Rearranging an algebraic equation is expected of a 'C' grade maths student, and one would expect that the majority of students would be able to do this. Not teaching them how to rearrange an equation will make it harder for the students to apply their knowledge to new situations later on, and they will struggle when the learning becomes more complex. For example, students using equation triangles to calculate kinetic or gravitational potential energy often make errors because there are four variables in the equation rather than the standard three. In addition, an equation triangle results in something else a student has to rote learn, which can be no help at all for some.

In short, they are worth a shot with *but* only at the right time – that is, once you have already taught the students how to use the equations mathematically (using the techniques below) and they have practised this. As some students don't love maths, they may find them a revelation and, as such,

I'd recommend introducing equation triangles in the run-up to exams as an alternative arsenal for them to use.

How to teach rearranging

Here are two methods for rearranging, both of which are likely to be taught by your maths department. The starting point is the same for both methods: you isolate the unknown before you rearrange the equation.

For example: Farida lifts a laptop 1.5m into the air. If she did 90J of work, how much force did she use?

Before you start, you must work out the unknown. First, break down the worded question. Ask the students to digest the information they know and write this down:

distance   = 1.5m

work done = 90J

The reason for this becomes clearer once they start answering more challenging questions. It also helps them to select the right equation to begin with.

The equation that needs to be used is:

work done = force x distance

Once the correct equation has been selected, I ask the students to start by highlighting the unknown in the equation. Then you can use one of the following methods.

Method 1: The balance method. To isolate force, you need to remove distance from the right hand side of the equation. Because distance is multiplied here, you can divide by distance to remove it. As you have to do the same to both sides of the equation, you also have to divide work done by distance as well:

$$\frac{work\ done}{distance} = \frac{force \times distance}{distance}$$

As distance is both multiplied and divided on the right hand side, the distances cancel one another out leading to:

$$\frac{work\ done}{distance} = force$$

Now that force is isolated, the numbers can be substituted into the equation (again, teach using the same vocabulary a maths teacher would use).

Method 2: Float and ping. Another method used by maths teachers is 'float and ping'. Whilst this is not as good at helping students to understand why they have used the method, it can be a useful technique for those individuals who struggle with numeracy. Also, some students just get it one way and not another.

The first step is to put a circle around the part of the equation you would like to remove:

work done $=$ force x (distance)

The circled part can then 'float' over the equals sign.

When the circled symbol 'floats' over the equals sign, there is a 'ping'. This is a reminder to students to change its sign to the opposite sign. So, in this example, the 'multiply by distance' becomes 'divide by distance'.

$$\frac{\text{work done}}{\text{(distance)}} = \text{force}$$

Getting students to remember physics equations

Students remember things when they understand them, so the first step should be to ensure that the science behind the equations is sound and that the students understand how the formula has been derived. However, once this has been established, it is worthwhile making sure that the students can recall them quickly as they will need to be able to do this in the exam.

Here are some good ways of helping them to remember:

- The equation dance. Get groups of students to make up an equation dance to help them remember the equation, and then have a competition in front of the class to see who has come up with the best one! Revisit these whenever the equation comes up in a question by getting them up on their feet and doing the dance.

| Force | Equals | Mass | Times | Acceleration |

- Mnemonics. A simpler version of the equation dance – for example, vets irritate rabbits ($V = I \times R$).

- Pictionary/Dingbats. What equation would this be?

- The equations race. This is an engaging way of getting the students to practise all of the above and is best used once they have been taught a number of different equations. Give the students a set of around 10 questions, using a variety of equations and including some tricky bits (e.g. rearranging, converting between units, different decimal places). Put the questions upside down on their desks. Insist on silence and then shout, 'Get ready ... set ... go!' When the students have finished, they can come to the front for their work to be checked by you. Give one mark for correct working, one for the correct answer and one for the correct unit (it helps to train them not to make mistakes). The first five or six students up to the front will not have full marks, so send them back. No one has completed the activity until they have achieved perfection.

Solving maths and science based problems

Teaching problem solving is often seen as a difficult topic. But whilst it can take time, and whilst progress is not always immediate, it can be taught really well. When topics are taught badly, students learn rules and how to use them in a straightforward setting. Teaching for problem solving requires them to think and see outside a set of rules, but this can be a big ask for students because, not only do they need to be able to synthesise previous knowledge, but they also have to apply it to a new context they won't have seen before.

The best way to teach problem solving is to model it as often as possible, demonstrating your thinking as you go through the problem. Some students will be able to do this 'thinking' without realising they are even doing it; others will need a bit more support.

A method for teaching this way is to teach students a mental plan of action. Here is a typical maths/science based problem that students may come across:

Describe and explain how the changes in the trend for smoking shown in the graph may have affected the occurrence of lung cancer between 1900 and 1980.

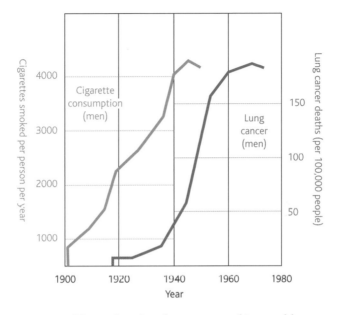

20 year lag time between smoking and lung cancer

This kind of question would generally elicit a very detailed response from a student who has been trained to think about what the data mean and link this to the science they need to learn. To many students, however, it would most likely result in a brief answer along the lines of:

> When the cigarette consumption went up, so did the number of deaths from lung cancer. Cigarettes therefore cause lung cancer.

To get the students into the correct mindset for answering these kinds of questions, I take them through a series of prompts. The aim is to use the prompts to teach the metacognitive processes required and then to slowly remove them so the students can think more creatively about their answers.

One way of doing this is to get them to RHASP!

- **R**ead once more. This might sound obvious, but when I refer to 'read' I also mean the graph so, at this point, I would get the students to annotate the graph like so:

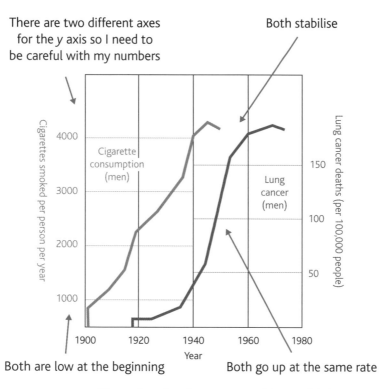

There are two different axes for the y axis so I need to be careful with my numbers

Both stabilise

Both are low at the beginning

Both go up at the same rate

20 year lag time between smoking and lung cancer

- **H**ighlight. The students should then highlight the key words in the question (describe, explain, the dates). This stops them from only describing the graph, only explaining the graph or, more commonly, only describing the main part of the graph rather than all the years that are asked for in the question.

- **A**rrange. This is the (so often neglected) planning section. The students can plan this out however they like but most prefer to use bullet points.

  They should first describe the graph:

  » The beginning – in 1900 the number of cigarettes per year and the number of instances of lung cancer are almost zero.

  » The middle – as the number of cigarettes consumed rises so does the incidence of lung cancer.

  » The end – as cigarette consumption stops increasing so does the rate of lung cancer.

- **S**cience. The students should list the scientific key words (i.e. cigarettes, tar, carcinogenic) that need to be included in the answer and then write their answer.

- **P**roofread. Finally, they should proofread their answer, checking that all of the bullet points have been included.

Once you have gone through the RHASP model with the students, get them into groups to work on a number of similar problems to cement this learning. There is no need to use this exact model; in fact, it's a good idea to come up with one that both the maths and science teachers can use at your school. The model should only be a starting point for teaching thinking skills, and hopefully you won't have to use it for too long as the students begin to assimilate the necessary skills.

Making maths a daily event

As with all of the ideas in this book, the crucial point is to make sure you do them all frequently in your lessons. The same goes for using maths. Do not leave it until it is a necessary part of a scheme of work or specification. The smallest use of it will normalise techniques and cement knowledge.

Here are some maths questions I've put to my students recently:

- How was the 50% Ebola virus death rate figure calculated in the 2014 outbreak?

- What fraction of you is … (insert carbon, William the Conqueror, etc.)?

- What is a nanobot? How small is it really?

- What is the probability of an asteroid hitting the Earth in our lifetime?

GETTING INSPIRATION

This chapter has not covered an exhaustive list of difficult topics, so it may be that your most hated topic is not here (sorry!). But there is a huge amount out there to help you to bring these topics to life.

When getting to grips with these difficult topics and areas you can do worse than search the Internet to find ways of making them engaging and furthering your own understanding of how to explain them clearly to students. Here are some websites you might find useful:

- http://ed.ted.com/lessons – TEDEd provides some brilliant resources to teach some pretty big ideas with great videos that explain concepts clearly.

- www.khanacademy.org – I would not advise using the Khan Academy site to teach students (particularly younger ones) but the video tutorials are well thought out and may help you to develop an understanding of how to explain difficult topics to your students.

- www.scienceinschool.org – This site includes some really good ways of making science fun and relevant.

- www.scibermonkey.org – Essentially a bank of links to websites with a variety of different ways to teach each topic. If you need inspiration, you will probably find it here.

Chapter 7
A GUIDE TO PRACTICAL WORK

'ARE WE DOING A PRACTICAL TODAY?'

I learnt almost everything I know about running an experiment, not from a fellow science teacher, but from the superb food technology team at my school. In my first term of teaching, my mentor sent me to see how practical work was taught in other subjects and I was in awe of the food tech teacher from the start. Every student in the school, from Year 7 upwards, used exactly the same 'hand claw' to cut vegetables, washed and cleared up like army cadets and always achieved the desired outcome at the end of every lesson. The contrast to my own first practical lessons was vast: low points included students jumping out of the window to get out of cleaning up and students using distilled water bottles to have a water fight. The worst was when a student desperate to hide their mistake from me managed to set fire to the bin, resulting in the fire alarm going off and the whole school being evacuated.

Determined to not repeat any of these experiences, I observed and probed the food technology teacher to learn from her mastery of the craft. There was consistency between members of the department in the way they did practicals, as if every instruction had a strict script to follow. So, regardless of which teacher had taught you by Year 9, at the start of GCSEs all the students followed the same procedures without having to be told what to do. This kept instructions to a minimum and focused staff and students on evaluating their work and improving their technique. It was no surprise that the food technology team were one of the most successful departments in the school with over 90% of students achieving expected progress in public examinations.

The lessons I learnt from them stay with me now. I am an advocate of practical work – even in April, right before the exams, I'm still to be found running through practical work with Year 11s. Some science teachers will question its usefulness, quoting evidence that it doesn't help students' learning.[1] Whilst it's true that poorly practised or disorganised practical work will not support students' learning, the right selection of tasks, coupled with probing questioning and skill development, will allow students to develop in so many more ways than simply understanding how to test for carbon dioxide.

And if you don't believe me, then listen to the Association for Science Education (ASE) who say that practical work is 'key to effective science attainment';[2] the Royal Society, whose advice is to increase both time and money investment in practical work;[3] and Ofsted, who in a survey of schools found those schools which had shown a clear improvement in their science departments used more practical lessons and focused on the development of skills.[4] Beyond the classroom, scientists are practical in their work and science is about discovery, so, to my mind, it is impossible to remove the experiment from our curriculum – and nor would we want to. But we do have to be careful to avoid what we might term 'the cookbook lab', where the students are given a 'recipe' and asked to complete it, without linking the practical activity to any scientific thinking.

When should you do practical work? I think the answer to this question is, whenever you can demonstrate a scientific phenomenon in the lab. This does not mean that I think students should have to do every science practical in the book. Sometimes practical work can be as simple as using a few

---

1   For a review of the research see: I. Abrahams and R. Millar, 'Does Practical Work Really Work?', *International Journal of Science Education* 30(14) (2008): 1945–1969.

2   See http://www.ase.org.uk/resources/secondary-science/.

3   See https://royalsociety.org/topics-policy/projects/vision/science-to-18/.

4   Ofsted, *Maintaining Curiosity: A Survey Into Science Education in Schools*. Ref: 130135 (London: Ofsted, 2013). Available at: https://www.gov.uk/government/publications/maintaining-curiosity-a-survey-into-science-education-in-schools.

students at the front of the class jumping up and down to demonstrate particle physics or a three minute teacher demonstration. The first question when thinking about practical work is, how are they going to learn this best?

DEMONSTRATIONS

First, consider whether a demonstration or a class practical is the best way to ensure understanding. Often a demonstration is beneficial in areas where there are student misconceptions, or when a practical would be technically difficult to perform or when it would take the students such a long time to complete that it results in little or no time left to analyse and understand the results. There are also some practicals that, due to health and safety rules, can only be completed by the teacher. The best of these, in student eyes at least, are those that involve explosions, such as thermite, and colourful hydrogen balloons.

Second, once you have decided that a subject would benefit from a demonstration rather than a practical, ensure that you have everything to hand. When you have 30 students huddled around the front bench, you do not want them to have to wait whilst you fumble about trying to find your matches. A good relationship with your technicians goes a long way here.

Finally, make sure you really think about how the demonstration is going to help the students to learn. Remember, magic is science without explanation; therefore, whenever you do a demonstration it really needs to be embedded in scientific theory.

Here are seven rules for how to complete an effective demonstration.

Rule 1: Be organised

If your demo is to go off without a hitch, you must plan it well and, especially if you are a new teacher, practise it first. This planning doesn't just mean the practical itself but also the questions you intend to ask to ensure the students understand the concepts. Below are some good examples of when and why a demonstration is useful and should be used.[5] [6]

pH rainbow in a burette	After practical work on acid and alkalis, use this to check that students understand the pH continuum. Set it up as a starter in the next lesson and get the students to discuss how the teacher has made it (students always remember colourful things).
Screaming jelly babies[5]	Use during the topic of respiration to help students understand that it is not breathing but a chemical reaction.
Burning a £5 note (or a £10 note if you're feeling fancy or flush)[6]	Use at the beginning of a topic on energy. Why does the note not burn? (In case you don't know this one, it's because you have previously soaked it in a mixture of ethanol and water and the water acts to keep the temperature around the note low enough so that it doesn't burn.)

5   See http://www.rsc.org/learn-chemistry/resource/res00000750/
      the-howling-screaming-jelly-baby?cmpid=CMP00000828.
6   See http://www.rsc.org/learn-chemistry/resource/res00000836/money-to-burn.

Electrolysis of water	This can be a really good way of demonstrating electrolysis and is great before the students complete a safer version themselves. It's best if you use a Hofmann voltameter or electrolysis cell so you can demonstrate that double the amount of hydrogen is formed (which can convince students about balancing equations).
Brownian motion	This experiment enables students to visualise diffusion at the particle level and can make the abstract come to life. As mentioned in Chapter 5, students have many misconceptions about particles and a demonstration can allow quality discussion time, facilitated by you, to overcome these. In addition, it is also very hard for the students to set up this experiment correctly, so a class practical on Brownian motion is a guaranteed headache without much in the way of results.

Rule 2: Show off!

I recommend watching the Royal Institution Christmas Lectures for a lesson in how to stage a demonstration well. They use demonstrations in the most captivating ways and, like them, you might want to think about costumes, props and how you are going to set the scene. Even if a demonstration is perhaps not the most whizz-bang example, you can still make it exciting for the students. To quite a large degree, teaching is acting so by pretending to be surprised by the results and by using showmanship, you can enliven even the most stolid demo. The best way I've seen this done was in a lower ability Year 7 lesson where the teacher called flame tests 'Harry Potter wands' and dipped the wooden splints into various salts and heated them in a flame whilst wearing a witch's hat and nose. You are only limited by the scope of your imagination and your tolerance for embarrassment!

Rule 3: Use the students

You only have two hands, but in a class of 30 students there are 62 hands in the room – and you can use them! There are some practicals, such as alkali metals in water, where health and safety means that students can't do the work, but for many others it should be possible for the students to measure out the chemicals, hold the equipment and record the findings on the board. Plus, if you have a student who is not particularly good at staying quiet during a demonstration, it can be a really good way of keeping them focused. All of this should allow you to have your eyes on your students the whole time and to focus on the theory behind the demo and your questioning.

Rule 4: Discussion is key

With demos, it's easy to slip back into the comfy cardigan of the teacher-led discussion. Whilst questioning to elicit understanding during demos makes sense, it's worth considering whether it moves all of the students forward or if it only helps the handful of students of whom you ask questions. Use some of the questioning techniques outlined elsewhere in this book – think–pair–share and thinking time work really well here. A graphic organiser like the one below can also help the students to structure their thinking.

Rule 5: Keep it simple (stupid)

Simple is best in a demonstration. Otherwise, you can find yourself explaining every part of what you are doing, linking in different ideas from other areas of science, and still the class are completely disengaged after a few minutes. Disengagement does not refer only to that handful of students who find being quiet difficult, but to the much larger proportion of your class who just aren't listening. Plan for *exactly* what you want the students to learn and which parts of the theory are important. Try not to deviate from this and keep your discussion to the point.

Rule 6: Enjoyment is fine

A former colleague of mine was dealing with a Year 9 class who had developed a profound apathy for science and for learning itself, and who expected all of the subject knowledge to be delivered to them on a plate. She changed things by introducing a low-tech, practical *Braniac* style starter demo to each lesson that was completely unrelated to the lesson itself. A session may begin by using a needle to pop a balloon and then repeating the experiment with a needle covered in oil. Students would then think–pair–share about why the second balloon wouldn't pop. What a simple way to develop students' thinking over time and to engage them back in a subject without taking up masses of valuable time.

However, there are times when, even if the students learn little more doing the practical than from a demo, you should still do the practical. Every student has the right to put a Mentos in a Diet Coke bottle simply because it is fun, and making honeycomb when studying reactions will help them to remember some of the basics in this area for years to come. Just try to balance the memorability of lessons with the time it takes to create them.

Rule 7: We all learn from our mistakes

If it doesn't go to plan, have a think about why it didn't work and make changes the next time. But move on quickly and don't dwell on your mistakes.

THE CLASS PRACTICAL

Almost every day of my working life, as I stand at the door welcoming my class with a chirpy 'Good morning', there are four boys who I can guarantee will ask, 'Are we doing a practical today, Miss?' This is before they have even considered whether it might be polite to say 'Hello'. Such is the seductive power of the practical! It could be that practical lessons are seen as being less cognitively demanding than theory lessons, but we are teaching science to develop the scientists of tomorrow and the skills they learn in practicals will be needed in future by those who take the subject further. Without using scientific enquiry in the form of student focused practicals, young people cannot develop their understanding of how new concepts are formed and old ones are disregarded. Some experiments will be flops and, in some, the students will have widely different results from one another. This can be used to illustrate the fact that, in reality, science can be quite a messy business.

What I learnt from the food technology teacher comes into play when you consider how to run a class practical. It really helps if there is a departmental plan or policy on how to run these, so if you pick up a new class in Year 10 you don't need to teach them how to go about doing a practical as you know that your colleagues set up experiments in the same way you do. If your school does not have such a plan, however, you may need to establish your own rules and routines.

The following section is purposely very instructional and is not directly linked to student learning. Introducing some simple procedures into your practical lessons will free you up to circulate and challenge the students on their results and develop their skills through tactful questioning when, otherwise, you might be tied up with 'practical matters'.

Establishing the practical lab

The first point to consider is the lab you teach in (this might need to be a more general plan if you teach in more than one lab) and how you will utilise all the available space.

Imagine the lab below:

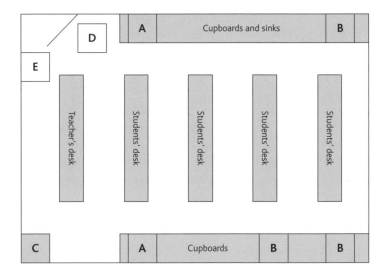

Within this, there are some areas I would advise you have set up as routine in your lab:

- A tidy-up station (A) to which students return dirty equipment and where they collect cleaning materials. Use two separate areas with large classes.

- Three areas in separate parts of the room (B) where you place any equipment the students will need, so you do not end up with students crowded around one area. Give the students a number from 1 to 3 and get those who are numbered 1 to go to equipment section 1 and so on. This will enable everyone to collect equipment quickly and smoothly.

- A safety equipment area (C) (glass bin, fire extinguisher, fire towel, etc.) which should be easily accessible.

- A clear space near the door (E) for health and safety for easy access.

- A display outlining the lab rules (D) (not just the standard 'no running' but also instructions on how to tidy up and what to do if something goes wrong).

- Labelled cupboards so the students can find generic equipment easily.

Right from the off with Year 7, aim to teach the students how exactly you perform an experiment – from listening to instructions to clearing away safely. This will save you from having to answer all those irritating 'Where is the thingy?' and 'Where should I tidy this?' questions and will, instead, allow you to focus on questions that really matter, such as 'Are these the results you expected?' and 'How can you explain that anomaly?' Use the same locations in every lesson and establish routines about how students tidy equipment away.

A note on health and safety

Never assume that students understand how to keep themselves safe, because they do not! Here are the top mistakes students will make:

- Not keeping their goggles on when pouring acids in a measuring cylinder near their eyes.

- Forgetting that equipment will be hot after it has been heated by a Bunsen burner (this applies especially to tripods) and assuming it will be okay to touch them.

- Putting on the gas before lighting the Bunsen burner, then walking around the lab to find a splint, have a chat, find a way to light the splint and, eventually, returning to the Bunsen burner, by which point it has spewed a fair amount of methane into the classroom.

- Hiding mistakes. I've observed students break some equipment, make a guilty face and then surreptitiously attempt to hide the evidence in the bin or the sink. To combat this you'll need to create an atmosphere where students are not scared of making errors (that is, provided it was an error and not a purposeful act of vandalism). Equipment will break. As a matter of course, you should make sure that students always use a test tube rack and, when something is expensive, make sure you tell them to be extra careful with it.

- Making intentional mistakes. In almost every class I have ever taught there has been a 'James'. James is fascinated by science and particularly by fire, explosions and bubbles. If you give him the opportunity, he will get all of the chemicals you have painstakingly laid out and, ignoring the safety instructions, will mix them all together to make bubbles, lots of bubbles, bubbles everywhere. Make sure he is sat at the front with a sensible student.

In addition to this, it is important that you work alongside your best friends (the science technicians) to make sure that you are clear on the Consortium of Local Education Authorities for the Provision of Science Services (CLEAPSS) guidelines for each of your practicals.[7]

Questioning

Getting routines established for practicals frees you up from spending time explaining to the students how to complete the practical and how to be safe, so you can use your time to support their learning instead. A poorly planned practical will result in conversations revolving around, 'No, *this* is the way you need to do it', 'Oh, I forgot to tell you all to put everything in a test tube rack' or 'Your results should look like this ...' When students are clear on how to complete a practical, where the necessary equipment can be found and how to be safe, questioning can focus on the experimental findings and the science behind the practical. This will allow you to ask things like, 'Can you explain to me why you are adding the salt solution at this point?' and 'Look at the results at Abdul's table – why do you think their results are different from yours?' In your planning, plot out exactly what you want students to think about and use this to help you devise your questions. Write these out in your plan to help you remember to ask them; if you just intend to make them up as you go along they may get forgotten (particularly if you are new to the profession).

Your instructions at the beginning of a practical are, of course, vital in ensuring that the students know what they are doing. But it can be very easy to have a classroom full of busy looking students who are learning zilch. This mainly happens when the students are just given a practical and asked to get

---

7   See www.cleapss.org.uk.

on with it. Don't tell the students what they are going to find out; instead, get them to think about why they are doing each step and what they think they are going to find out and why before they start the practical. This might seem like a big ask, especially if you only have a 50 minute lesson, but if time is a real problem then set up the practical in the previous lesson.

Grouping students

Most of your practicals will take place in groups or pairs. This tends to be the best idea, not just because of the amount of equipment required but because it provides more hands with which to get a task going. In addition, the students learn a great deal from dialogue with one another, but you'll need to orchestrate this if it is going to have the utmost benefit (e.g. by structuring think–pair–share activities, by getting pairs of students to question one another's results, by providing prompt questions to support their discussions). Group work is also a really easy way to differentiate for the differing needs of your students, although there are advantages and disadvantages depending on how you organise the groups (as summarised in the table below).

Groupings in practical work	
The advantages of ability groupings	**The advantages of mixed ability groupings**
Groups can work on slightly different tasks (e.g. one group can study the effect of temperature and pH on enzymes and another can look at just pH).	This way of organising the groupings can allow some students to shine if their practical skills are stronger than their ability to show it on paper.

Differentiation can be a breeze: step-by-step instructions can be given to support some groups whilst general guidance is given to others.	Lead learners can develop others through guiding and helping.
Questioning can be specific to the progress of the group.	Students can regulate one another's behaviour.

This said, if you have ever taught a Year 12 A level science class and tried to get students to complete practicals independently, then you may well have been surprised at their inability to work alone. So, as well as thinking about ability vs. mixed groupings, look out for opportunities for students to work on their own (though your technicians may not like this at all!).

Getting the students to think

Obviously, getting students to think is what you want practical work to achieve, but it is well worth asking yourself what *exactly* you want them to think about. The scientific method is not a simple thing, but Year 7 students are often taught straightaway to predict, write a method, make observations, form conclusions, link them to the science and then evaluate. That's a hell of a lot for an 11-year-old to take in. You should take a Year 7 group through this process slowly, with one practical focusing on prediction, another on evaluation and so on until they have a basic understanding of what completing a practical actually means. Then continue with this focused approach, as it can become apparent that a Year 11 student may well grasp the first half of the method (predicting and making tables) but may still need a lot of work on the second half (ensuring the conclusion is embedded in science and using vocabulary such as 'concordant results', 'valid' and 'reliable' in their evaluation).

If you are doing a straightforward, simple practical then teaching students to go through POE (predict, observe, explain) as a routine can be a quick way of making sure they have thought about it. But, as a method, it is unlikely to push them any further in their scientific skills. Instead, by choosing one element, you can focus on mastery of that area and then use this to give detailed feedback.

Assessment and feedback

Assessment and feedback has been a hot topic for science teachers lately: should we or should we not assess practical work in our schools? Well ... even if the powers that be decide not to assess practicals at exam level, evaluating practical work in a classroom setting has genuine importance. Without assessment (and by this I do not just mean giving the students a grade), progress in the practical classroom will not happen and our students will not be ready for studying the subject further at A level or degree level. The first step is ask yourself what you want to give feedback on. This goes back to rule 1 above about being organised, so we need to think about this before the lesson starts. Writing lab reports can be a really good tool, but they can also be time consuming, particularly if you want to give the students time to redraft. Instead, assess and give feedback on one particular area for each practical and then get this redrafted. You will see much better results. This is a place where model answers can really come into their own for showing students what is expected of them (more on this in Chapter 8).

Alternative lab reports

Technical writing is a skill that we need to teach the students, but occasionally one of the following options can make it seem a bit more exciting:

- Get the students to write a news article in character as a scientist whom no one believes. Alternatively, ask them to write in character as a scientist who is showing off to rival scientists about how they got to a breakthrough first.

- Use this breaking news tool to summarise a group's findings (it only takes a second): www.classtools.net/breakingnews/.

- News flash. Record the first group to finish by doing a news flash on what they found out. Share it as part of your plenary to see if the rest of the class agree with their findings.

- Twitter lab report. Get the students to condense their conclusion into 140 characters. Then publish it!

- Use Google Docs to collate the results of all groups and use this to form the conclusions of the class. This focuses on the collaborative nature of science and the requirement for data to be reliable.

- Try the wordless essay (see Chapter 3).

- Ask the students to explain the practical to a Year 4 student. This teaches older students to be clear and concise.

- Consider keeping student practical portfolios. This can be a really nice way to show the students how much better they are getting at writing practical reports.

In summary, make sure you are running a proper practical and not falling into the traps of the 'cookbook lab'. In a proper practical:

- The topic is introduced and the students develop their own questions.

- The students make predictions.

- The students develop the plan and identify the required materials.

- The students decide how to collect data and how much data is good enough.

- The students analyse results and use secondary data to support their conclusions.

- The students evaluate and question in detail what they have found out.

- The teacher is there to feed back to the students and to probe with challenging questions.

In the cookbook lab:

- The theory is explained first.

- Routines are not in place.

- Questions are given to the students at the beginning.

- A method and detailed plan are provided.

- A data table is given.

- The students work through the plan independently.

- The teacher focuses on practical issues during the experiment.

- The lesson runs out of time and conclusions are not made by the students.

Chapter 8
MOVING STUDENTS FORWARD

'I GOT THE SAME GRADE – AGAIN'

When I was an NQT, the one thing I really struggled to get to grips with was assessment. I marked books because I knew someone would check that I'd done so at some point and that I would probably have yet another observation just around the corner. I didn't do it because I thought it was useful, but kept on top of it regardless of feeling that it was making very little difference to the learning. My general marking looked a little like this:

But this kind of marking doesn't really help the students to understand what it is that they know well or what they should do to improve. What does 'add detail' mean to a student? Nothing really. It didn't take too long for me to change my mind and my practice.

Good assessment is essential to move skills forward and identify how to modify your teaching and planning, but assessment is not just about marking their books every two weeks or giving them a test at the end of the unit. It encompasses all of the following:

- Assessing the students from the beginning.

- Rigorous formative assessment with feedback that actually changes things.

- Analysing and closing the gap (i.e. helping students to get from what they currently know to where they need to be and then checking they have actually done this).

- Marking for literacy.

- The test!

The impact that good assessment and feedback can have on students' learning is immense. In Hattie's meta-analysis, the average effect size of giving specific feedback was 0.79 – twice the average effect size of most interventions.[1] But it is only this effective when feedback is done in the best way possible; tick and flick might have an effect size of near zero.

The main point is that we never, ever start teaching a topic by assuming the students have no prior knowledge, because if you presume this then you end up teaching Year 7 stuff they already know, and there is little point in that. The flip side is that, without ascertaining where the students are, you might assume they understand basic ideas that they in fact don't and have to go all the way back to the basics after you have pitched a lesson too high.

---

1   Hattie, *Visible Learning*.

THE PRE-ASSESSMENT

This is an idea that my current science department do very well indeed: a pre-assessment is a formalised way of understanding how much a student already knows about a topic before you start teaching it. Some teachers present this as a checklist which they can then refer to at the end of the unit. This is nice for showing progress but sometimes students don't understand their learning, are unaware of their misconceptions and, sometimes, they just don't take it very seriously. This is where a short low stakes quiz can be a really useful tool. The quiz should focus on:

- What you expect students to be able to do from previous learning (i.e. primary school, previous units).

- What you are going to teach them.

- Any common misconceptions you are aware of around this topic.

Some open questions are needed to find out what they know from previous topics so you can have an idea about their general understanding. These should be set alongside multiple choice questions to check for more specific understanding of certain areas or misconceptions. You will need to encourage the right kind of classroom culture so the students understand that knowing their weaknesses is how they get better. The quiz should last no more than 10 minutes.

Here is an example used before a Year 7 topic on cells, variation and evolution.

1   What do living things need to survive?

2   Give four examples of variation in dogs.

3   What do mammals have in common?

4   What makes up our bodies?

5   Name three differences between animals and plants.

6   Draw a food chain with at least three organisms in it.

7   Where does the energy originally come from in your food chain?

8   What is DNA?

    a   It is found in your cells and makes you who you are.

    b   It is an alkali.

    c   It is in your brain.

    d   It is part of a food chain.

9   What is evolution?

    a   Chimpanzees are our ancestors.

    b   How animals change over time.

    c   A change in characteristics over generations.

    d   That we are all different to one another.

10   Who came up with the theory of evolution?

    a   Rutherford

    b   Darwin

    c   Fleming

    d   Watson and Crick

One of the issues with a task like this is that there may be students who can't answer many questions at all, so this is where the all-pervading supportive culture you should have in your classroom is important. To create this environment, ensure that students understand prior to the assessment why you are doing it – that is, to help you to know what they know and what they don't, to help you to plan the best lessons and not so that you can judge/grade them. At the end of the topic, get them to look back at the pre-assessment and see how much progress they have made. If you commit to using a pre-assessment they will become routine and students will not mind not being able to answer all of it.

Finally, using a pre-assessment is useless unless it is then used to inform your planning. It is the ultimate way to differentiate a topic as you should now have a really good idea of where to start. If you don't like the idea of starting a new topic in this way then it is possible to use any kind of quiz or questioning instead, but it is unlikely to give you so much useful information.

SETTING THE SCENE

Once you have a good idea of the strengths and the gaps in your students' understanding, the next step in providing good feedback and assessment is to ensure the students truly understand what they need to be able to do before they start a task or series of tasks. The aim is to establish what good and excellent looks like, otherwise how will they know? These may be called 'success criteria' in your school or something else, but whatever they are called they must be specific to the task, be easily understood by all and still be challenging enough to push every student beyond what they think is their best. This is an opportunity to build the culture of achieving your personal best and always aiming to improve.

Level ladders or different structures can be used here, but the most success-ful method is to show the students the work of previous students, using these to set the scene for each level.

This would be an okay example of an explanation of the digestive system:

> The digestive system is the system of organs which digest your food so that you have the energy and the nutrients you need to live and be healthy. First, the mouth chews the food, which moves down the oesophagus to the stomach and small intestine, where the food is absorbed to the large intestine and then to the anus.

And here would be an excellent (although maybe not perfect) one:

> The human digestive system is an organ system which is responsible for breaking down the large molecules in your food into smaller mol-ecules that can be absorbed into the blood system and used to pro-vide energy and nutrients. To start with, the mouth chews the food into a bolus and adds digestive enzymes to the food so that digestion begins immediately. The food moves through peristalsis down the oesophagus to the stomach. From the stomach the bolus moves to the small intestine where further digestive enzymes are added result-ing in small molecules such as glucose and amino acids which can then be absorbed through the villi in the intestine walls.

Exemplar work can be provided throughout the lesson and should be dis-cussed with the students. What was good about the first example? What could be done to improve it? At this point break down the answer into its various areas. What is good about the introduction? What about the explanation of the role of the mouth? What about the explanation of the stomach? Taking photos of students' work mid lesson and then displaying it on an interactive whiteboard is a much better way of identifying strengths and weaknesses than asking students to read their answers out loud, as it gives you the opportunity to annotate their work with your class.

Part of this process comes down to how much you understand the grading criteria and how well you communicate it to the students. This is not easy, as the criteria in science are really hazy (and most grade criteria are lost on students anyway). Is understanding protein synthesis really an 'A' grade topic as most grade descriptors have it? I would expect a 'C' grade student to know elements of this (including the two steps and what its purpose is) and for an 'A' grade student to be able to fluidly describe and link the necessary stages. The best departments I know have grade descriptors for each topic outlined with examples of student work for each grade, which is especially helpful if you are an NQT.

Before you begin teaching is also the time when you should plan how you are going to assess the students. Planning a curriculum for what you really want it to include is useless unless you define how to assess how well a student achieves that goal. If the curriculum is developed but the assessments do not reflect the change towards skill development or hinge concepts, then the teaching will not happen successfully enough. Teaching towards the test is inevitable, but make sure you are testing the students in a way that ensures the curriculum is how you want it to be. Some questions to consider here are:

- How will skills be assessed throughout the unit?

- Now that you are interleaving your topics, how can this become part of your assessments in order that they test more than the current topic?

- How will you use assessment to make a difference to students' learning?

FORMATIVE ASSESSMENT

A Wellcome Trust survey into students' perceptions of science revealed that they were critical of the way they were assessed, feeling there was an over-reliance on tests and exams.[2] Tests can be useful, of course (and more on summative assessment later), but employing a variety of assessment methods was the main recommendation of the report.

Assessment is formative if it helps the students to know what they need to do better and how they might do this. Thinking back to my earlier example of poor marking, it's very unlikely that this would have helped the student receiving it in any way. A quick literacy ninja quiz at the end of the week might have been a more useful opportunity for the student to identify what they needed to work on. If, for instance, a number of students have struggled with the definition for osmosis, make sure that in the next ninja test this word is included again (and again the next week or until they've got it). Additionally, assessment does not need to be just a dull test – you could include osmosis in a game of bingo or splat to reinforce it.

To create a culture where making mistakes is seen as a learning tool, it is important that most assessments are low stake. A really easy way to achieve this is to not give grades on all assessments. There are a number of reasons for this:

- Grading work results in students working hard for a grade or a goal rather than for the sake of learning something new or mastering a skill.

- Grades reinforce the idea that you must strive to get the correct answers. This goes against a culture where mistakes or misunderstandings are seen as a way to learn.

---

2   National Foundation for Educational Research, *Exploring Young People's Views on Science Education*.

- Grades create a fear of failure (even in the most able students).

- Grades allow students to see where they fit when ranked alongside their peers but provide no support on their own on how to improve.

- If you give the students a grade and comments on how to improve, they are likely to focus on the grade and completely ignore the comments.

Assessment also needs to be regular – the act of retrieving knowledge during a quiz is one of the most useful in cementing long term memory by providing an opportunity for repeated practice, in addition to acting as a means of receiving valuable feedback. Using a variety of methods, not just quizzing, will result in effective practice and skills development.

Improving students' skills in science will also require detailed and individualised feedback if they are to grasp how to move forward – the kind of feedback that will not be possible through simple testing. A quiz can check that the students know the definition for interdependence but it will not test their understanding of its role within a given ecosystem. For that, the students will need an opportunity to look at some data (e.g. the impact of an increasing snow leopard population on numbers of hares) and explain it. In this example, students often offer simple explanations of what may be happening (e.g. 'There are more predators so the number of hares reduce'), but they can't predict what might happen eventually to the population of snow leopards once hare numbers have reduced significantly. This is where feedback can really help.

Assessment must be linked to giving feedback to students that provides them with a clear idea of how to move forward – and this is the key to good marking.

BETTER MARKING

The number one principle to remember when marking is that assessment must enable a student to know what they need to do to improve and then get them to do something about it. Simply telling students what broad area they need to work on will only get them so far. And how will it help those students who are not perhaps the best learners yet? Those slightly lazy ones at the back who, when told they need to work on their descriptions of how a nuclear power station works, simply read your feedback, ignore it and move on.

For these students, the simple WWW (what went well) and EBI (even better if) model is defunct as it cannot give them the confidence required to do any better. The WWW/EBI mode of marking is often a tick box exercise aimed at keeping the head of department off your back, but often because WWW/EBI is used as part of a two week marking style you may be going through the motions and giving an EBI for something that they did five or six lessons ago – something they can't even remember having written. This kind of marking can end up being a useless shortcut because the students cannot engage with its meaning and so it does nothing to move them forward.

The key to effective marking is that the feedback given must be constructed in a manner that will support students as they try to move on to the next level in their learning. The feedback needs to give them a reason to improve their work, bridging the gap between what a student needs to know and what they know at the moment (harking back to the ladder in Chapter 2). For the feedback to do this, it should provide students with the necessary confidence, it should focus on the task (not the student), it should be manageable and chunked, and it should include all the detail the student needs to get there.

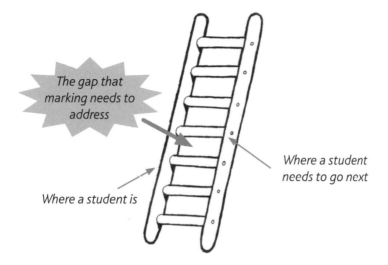

The gap that marking needs to address

Where a student needs to go next

Where a student is

So, to return to my original example at the start of this chapter, does it meet this criteria? The answer is clearly no.

A better example of useful feedback would be:

In your answer you have correctly identified that NaCl is an ionic compound and its properties. To move this further:

1   Draw a dot-cross diagram to show formation of NaCl.

2   Explain why ionic compounds have those properties.

    a   High melting point because they are strong bonds.

195

b   Can conduct electricity when molten
    because the ions are free to move.

You have clearly put a lot of effort in here.

To improve the practical write-up you need to develop your evaluation.

1   You have correctly identified two improvements. Now you need to *explain* why this would make a difference.

2   In your explanation you need to refer to the scientific principles.

> You have clearly put a lot of effort in here
>
> To improve the practical work up you need to develop your evaluation.
>
> ① You have correctly identified two improvements. Now, you need to explain why this would make a difference
>
> ② In your explanation you need to refer to the scientific principals

One of the areas where science teachers often struggle with marking is when feedback is not needed to improve a student's skill but to improve their factual knowledge or understanding of a topic. The gap model can work just as well here, but in this case you will actually need to give the students something to do – in other words, a task that will support them to move forward. In the examples above, the teacher has set some further questions to scaffold the student's answer to improve an essay style question. Once the student has completed the task, the teacher has gone back to comment on a further way they can improve, thus not accepting work unless it is completed to the best of their ability.

For this kind of marking to work at its best you need to give sufficient time for it. Once you've marked their books, you need a good 15 minutes at the beginning of the lesson dedicated to a student feedback session where they are expected to respond to your marking. This might feel slow, but by doing this you are ensuring that the students close the gap before you move on to the next topic. Students are often at pains to get on to the next topic and can be reluctant to engage with feedback, but working in this way improves their skills and understanding. So, make sure you give it enough time and, eventually, they will realise how useful it is.

Here are some other ideas for marking:

- Do not accept work for marking until it is at an acceptable level – use self and peer assessment as a precursor to marking to make sure that the work you are giving feedback on is of the best quality.

- Use quick notations to convey common ideas. One teacher I know initials work in green pen if the quality of the student's work is improving/good quality or a red pen if it is not. Others use +, – or = signs to signify the same ideas.

- Put an asterisk in the margin of a student's work if there is an error and get them to work out (perhaps with a peer) what it is and what they need to do to get it right. You can also do this as you walk around a classroom when students are working on a task.

- A tip to save time when giving students a task to respond to is to write the questions on a PowerPoint slide alongside a number and then write a number in the student's book for the question they need to answer (as in the example below). Typically, you may end up with between three and five questions in total that the students need to address to move forward within each class, so this will save you having to write up the questions each time.

Feedback – Mind the gap

Answer the question identified in your marking to close the gap.

1. Explain how molecules are moved around the plan. Use examples of active transport, osmosis and diffusion.

2. Explain how static electricity can be dangerous when refuelling a plane.

3. Describe and explain the data from the experiment into CO_2 levels and photosynthesis.

CO_2 (in ml/l)	Number of bubbles of O_2 released by the plant (in 60 secs)
2	4
4	8
5	9
8	9

- You will not have time to mark every piece of work a student does, but you should definitely mark work that it is important for the students to do well. If they are struggling on analysing data and describing graphs, then mark this section of their work rather than the hypothesis they made for the experiment.

- Always mark for literacy. It is not just an English teacher's job! If other teachers don't do it then the students will not see it as important, so don't ignore spelling mistakes.

- There is real power in the 'pre-mark' task. This is when you ask the students to reflect on their learning before you collect in their books – I either get them to fill in some thinking bubbles or use a learning triangle like the one below.

1 thing you knew before the lesson

2 questions you have

3 things you learnt this lesson

SELF AND PEER ASSESSMENT IN SCIENCE

Psychologists Barbara White and John Frederiksen conducted a study looking at the importance of self and peer assessment on 12 classes of 30 physics students.[3] The students were studying force and motion and the classes were each split into a control group and an intervention group. The control group spent some lesson time discussing the work and the intervention group engaged in both self and peer assessment. The results of three subsequent tests showed a real improvement in the attainment of the intervention group, particularly amongst the weaker students. The conclusion was that students do not underachieve because they don't understand the work but because they don't know what knowledge to focus on.[4]

When self and peer assessment is used against really clear learning goals, it can make a difference to the student learning. My experience of both self and peer assessment is that they are useful when the learning is simple (e.g. remembering the role of each organ of the digestive system). They help students to strengthen their memory recall, therefore they are good for reinforcing understanding. However, when it comes to moving students forward with regards to skill development they fall short. If a student doesn't understand how to pick apart a graph, as opposed to simply explaining a general trend, how is self-assessment going to help them to do so? (It may provide a nudge if something has been forgotten but nothing more.) And if a student has their work peer assessed by another student who doesn't

---

3   B. White and J. Frederiksen, 'Inquiry, Modelling, and Metacognition: Making Science Accessible to All Students', *Cognition and Instruction* 16(1) (1998): 3–118.

4   For a round-up of the data on peer and self-assessment see: J. Sebba, R. D. Crick, G. Yu, H. Lawson, W. Harlen and K. Durant, *Systematic Review of Research Evidence of the Impact on Students in Secondary Schools of Self and Peer Assessment*. Technical Report no. 1614 (London: EPPI-Centre, Social Science Research Unit, Institute of Education, University of London, 2008).

understand a certain principle, how can they identify the relevant issues and give feedback?

A quick look in the books of my classes will show self and peer assessment employed for pop quizzes, vocabulary ninja and perhaps where a checklist is given (or a skill is being practised yet again – for example, in graph work with a checklist to remember axes, units, line of best fit, etc.). But where learning gains are really important (such as evaluating data to assess if it provides enough evidence to form a conclusion), then there is no substitute for individualised feedback from the teacher. Verbal feedback is great too – just don't use those awful verbal feedback stamps (what on earth is the point of those?!).

Self-assessment

Self-assessment can be useless if it is not done well, but there is real merit in getting the students to draft and then check their work before it is handed in to you. The best way to make redrafting really useful is to give very clear instructions to students on what to look out for in their work and then get them to highlight the parts of the text that is concerned with these areas. For example, during a project analysing rates of reaction, you could give the following instructions to help the students improve their work:

- Highlight the key words used in your text – there should be at least six of them.

- Highlight your definition of each key word in another colour – you should have at least six definitions.

- Put a red square around each of your particle drawings.

- Use a blue pen to asterisk the explanation of each drawing – there should be four stars.

This helps the students to identify what they've missed out, which means that when you mark their work you are giving them the feedback appropriate to their level of learning rather than on a section which they hadn't checked and in which they had therefore made errors.

Peer assessment

The late Graham Nuthall's research revealed that 80% of the feedback a student gets in a lesson is from their peers, and most of that is wrong.[5] This statistic might be employed to argue against the efficacy of peer assessment, but my feeling is that it just shows how careful you need to be when using it if you are to obtain the best outcomes.

The first step in peer assessment is to ensure that there are ground rules so students understand they need to be kind, specific and helpful.[6] Once this is set up, it is worth giving the students sentence starters and comment banks to support their answers.

Sentence starters	Comment bank
This work meets the criteria because …	You have drawn four free body diagrams.

5   G. Nuthall, *The Hidden Lives of Learners* (Wellington: NZCER Press, 2007).
6   See R. Berger, *An Ethic of Excellence: Building a Culture of Craftsmanship with Students* (Portsmouth, NH: Heinemann, 2003).

I like your work because ... One strength of the work is ... I like your explanation of ... One thing that you could improve is ... To improve the work you could further explain ... Could you use a diagram to demonstrate ...?	You have remembered to keep the arrows the same size on a balanced object. You have explained what happens when an object is moving at constant speed. You have come up with your own examples of free body diagrams. You have explained how forces change on a falling object. You have mentioned that force is measured in newtons. You have shown how to calculate resultant force.

The following strategies are easy ways to use peer and self-assessment:

- Exit pass. Use an exit pass of key questions that the students have to complete before they leave the lesson, then get them to self or peer assess their answers in the next lesson.

- Weekly summary. Do a quick quiz at the end of every week to check they have understood the week's lessons. This can be easily self or peer assessed, provided the focus is on knowledge and understanding rather than skills. Collect in the grades to help you identify which topics to spend more time on later. Prepare some 'do next' tasks for the students to complete based on the issues raised from the test.

- Vocabulary ninja (see Chapter 4) can certainly be marked by a peer. The answer sheets can be quickly handed in to you for a brief review to pick up on any common errors.

Self and peer assessment cannot be a substitute for expert teacher feedback, but it can help to dramatically reduce marking time and improve students' self-regulation and error checking.

A few ideas for creating a culture of trust in the classroom include:

- Promote mistakes as a learning tool – give examples of your own errors and stress that mistakes are welcome.

- Refer to first drafts as 'first attempts in learning'.

- Make redrafting part of your teaching process.

SUMMATIVE ASSESSMENT

Tests are useful – we've already seen the difference that regular quizzing can make to students' learning in terms of retrieval practice. Summative assessment usually refers to the end-of-term tests or exams that students complete to check how well they have done. I would argue, however, that no assessment should be purely summative. If a student has achieved 67% on a Year 7 test on space, then this means something – they will have shown strengths and some weaknesses too. How we use our data to analyse a student's performance is critical in helping them to move forward.

Data can be a minefield: ignore it and it's useless; use too much and the important findings are lost. But employ it well and it can really help students to make progress. Here are some ideas:

- Use a spreadsheet, highlighted in red, amber and green to analyse what questions the class did well and which ones they didn't. Spend a lesson going over these areas with a new slant to help support understanding.

- Get the students to complete an analysis of their own test to identify their strengths and weaknesses and then suggest tasks for them to complete.

- At the end of any half-term, leave one week before the holiday as a review week to ensure the students have mastered a topic before you start on the next one after the holidays.

- Ask the students to redraft important areas of the test.

For your summative assessments to be useful, they will need to assess the most important science in the topic that you are looking for the students to understand. Look through your assessments and ask yourself, what percentage are skills based and what percentage ask the students to apply their knowledge? Are there sufficient opportunities for them to write at length? Have you included questions that test their ability to use maths scientifically given how important this is becoming in GCSEs? Does the assessment allow you to not have a trade-off between preparing for the test and being able to teach those spontaneous moments where a student asks you an 'out there' question on the topic?

Finally, all of this assessment and feedback only works properly if the students work harder than you do: all the feedback in the world can be wasted if the students don't do anything with it to move themselves forward. Don't forget they are kids – they don't like criticism and they won't respond to it unless there is a culture of trust in your classroom.

AND FINALLY ...

Teaching is one of the most rewarding careers there is and it is also one of the most exhausting. This book has set out some ways to make teaching science more exciting, but don't plan lessons so meticulously that you lose your enthusiasm for the subject and the students. So much of teaching relies not on the content of what you teach but on your enthusiasm – the tone of your voice and your body language, on your relationships with those around you and how much you are able to convey that you care. Don't be afraid to make mistakes. As long as you haven't blown up the lab or allowed someone to electrocute themselves, then you're doing alright!

We are all learners and whilst I feel privileged to have written this book, I am still not an expert – so please do get in touch with any ideas or thoughts you want to share.

Oh, and the answer to the maths question on page 143–144 is (a).

@CatrinGreen

BIBLIOGRAPHY

Abrahams, I. and Millar, R. (2008). 'Does Practical Work Really Work?', *International Journal of Science Education* 30(14): 1945–1969.

Allain, R. (2014). 'How Do We Know the Earth Orbits the Sun?', *Wired* (14 April). Available at: http://www.wired.com/2014/04/how-do-we-know-the-earth-orbits-the-sun/.

Berger, R. (2003). *An Ethic of Excellence: Building a Culture of Craftsmanship with Students* (Portsmouth, NH: Heinemann).

Bergmann, J. (2012). 'Just How Small Is An Atom?' [video], *TED* (April). Available at: http://www.ted.com/talks/just_how_small_is_an_atom.

Biggs, J. and Collis, K. (1982). *Evaluating the Quality of Learning: The SOLO Taxonomy* (New York: Academic Press).

Bjork, E. L. and Bjork, R. A. (2011). 'Making Things Hard on Yourself, But in a Good Way: Creating Desirable Difficulties to Enhance Learning', in M. A. Gernsbacher, R. W. Pew, L. M. Hough and J. R. Pomerantz (eds), *Psychology and the Real World: Essays Illustrating Fundamental Contributions to Society* (New York: Worth Publishers), pp. 56–64.

Block, M. (2006). '"Voices from Chernobyl": Survivors' Stories', *NPR* (21 April). Available at: http://www.npr.org/2006/04/21/5355810/voices-of-chernobyl-survivors-stories.

Bohr, N. (1987). *The Philosophical Writings of Niels Bohr,* Vol. 2: *Essays 1932–1957 on Atomic Physics and Human Knowledge* (Woodbridge, CT: Ox Bow Press).

Boohan, R. (2016). 'The Language of Mathematics in Science', *School Science Review* 360 (March): 15–20. Available at: https://www.ase.org.uk/journals/school-science-review/2016/03/360/.

Brown, P. C., Roediger, H. L. and McDaniel, M. A. (2014). *Making It Stick: The Science of Successful Learning* (Cambridge, MA: Harvard University Press).

Bryson, B. (2003). *A Short History of Nearly Everything* (London: Black Swan).

Cooke, E. (2008). *Remember, Remember: Learn the Stuff You Thought You Never Could* (London: Viking).

Cox, B. (2012). 'A Crash Course in Particle Physics' (1 of 2) [video], *YouTube* (1 January). Available at: https://www.youtube.com/watch?v=HVxBdMxgVX0.

Darwin, E. (1794). *Zoonomia; or the Laws of Organic Life* (London: Johnson).

Dawkins, R. (ed.) (2009). *The Oxford Book of Modern Science Writing* (Oxford: Oxford University Press).

Dobzhansky, T. (1973). 'Nothing in Biology Makes Sense Except in the Light of Evolution', *American Biology Teacher* 35: 125–129.

Dunlosky, J., Rawson, K. A., Marsh, E. J., Nathan, M. J. and Willingham, D.T. (2013). 'Improving Students' Learning with Effective Learning Techniques: Promising Directions from Cognitive and Educational Psychology', *Psychological Science in the Public Interest* 14(1): 4–58.

Enard, W., Przeworski, M., Fisher, S. E., Lai, C. S. L., Wiebe, V., Kitano, T., Monaco, A. P. and Pääbo, S. (2002). 'Molecular Evolution of FOXP2: A Gene Involved in Speech and Language', *Nature* 418: 869–872.

Francis, M. (2007). 'The Impact of Drama on Pupils' Learning in Science', *School Science Review* 327 (December): 91–102. Available (for members of the Association for Science Education) at: https://www.ase.org.uk/journals/school-science-review/2007/12/327/.

Gray, R. (2008). 'Why Elephants Are Not So Long in the Tusk', *The Telegraph* (20 January). Available at: http://www.telegraph.co.uk/news/science/science-news/3322455/Why-elephants-are-not-so-long-in-the-tusk.html.

Gribbin, J. (2012 [1985]). *In Search of Schrödinger's Cat* (London: Black Swan).

Halikari, K., Katajavuori, N. and Lindblom-Ylänne, S. (2008). 'The Relevance of Prior Knowledge in Learning and Instructional Design', *American Journal of Pharmaceutical Education* 42(7): 712–720.

Harlen, W. (2011). *ASE Guide to Primary Science Education* (Hatfield: Association for Science Education).

Hattie, J. (2008). *Visible Learning: A Synthesis of Over 800 Meta-Analyses in Education* (Abingdon: Routledge).

Hattie, J. and Yates, G. (2013). *Visible Learning and the Science of How We Learn* (Abingdon: Routledge).

House of Lords (2006). *Science Teaching in Schools: Report with Evidence of the House of Lords Science and Technology Committee*. HL Paper 257 (London: The Stationery Office).

Linn, M. C. (2006). 'The Knowledge Integration Perspective on Learning and Instruction', in R. K. Sawyer (ed.), *The Cambridge Handbook of the Learning Sciences* (New York: Cambridge University Press), pp. 243–264.

McDaniel, M. A., Agarwal, P. K., Huelser, B. J., McDermott, K. B. and Roediger, H. L. (2011). 'Test-Enhanced Learning in a Middle School Science Classroom: The Effects of Quiz Frequency and Placement', *Journal of Educational Psychology* 103: 399–414.

McNamara, D. S. (2010). 'Strategies to Learn and Learn: Overcoming Learning by Consumption', *Medical Education* 44(4): 340–346.

Millar, R. (2005). *Teaching About Energy*. Department of Educational Studies Research Paper 2005/11 (York: University of York Department of Educational Studies).

Millar, R. and Osborne, J. F. (eds) (1998). *Beyond 2000: Science Education for the Future* (London: King's College London).

Mould, R. F. (2007). 'Pierre Curie, 1859–1906', *Current Oncology* 14(2): 74–82. Available at: http://www.ncbi.nlm.nih.gov/pmc/articles/PMC1891197/.

National Foundation for Educational Research (NFER) (2011). *Exploring Young People's Views on Science Education: Report to the Wellcome Trust*. Available at: http://www.wellcome.ac.uk/About-us/Publications/Reports/Education/WTVM052735.htm.

Needham, R. (2015). 'The Language of Maths in Science', *Royal Society of Chemistry Education in Chemistry Blog* (13 May). Available at: http://www.rsc.org/blogs/eic/2015/05/language-maths-science-graphs.

Nuthall, G. (2007). *The Hidden Lives of Learners* (Wellington: NZCER Press).

Ofsted (2013). *Maintaining Curiosity: A Survey Into Science Education in Schools*. Ref: 130135 (London: Ofsted). Available at: https://www.gov.uk/government/publications/maintaining-curiosity-a-survey-into-science-education-in-schools.

Orzel, C. (2010). *How to Teach Quantum Physics to Your Dog* (London: Oneworld).

Perlin, D. and Cohen, A. (2002). *The Complete Idiot's Guide to Dangerous Diseases and Epidemics* (Indianapolis, IN: Alpha).

Pessoa, L. (2008). 'On the Relationship Between Emotion and Cognition', *Nature Reviews Neuroscience* 9: 148–158.

Robinson, M. (2014). 'Classroom Practice: Don't Just Talk at Them, Spin a Ripping Yarn', *TES* (14 February). Available at: https://www.tes.com/article.aspx?storyCode=6403314.

Roediger, H. L. and Karpicke, J. D. (2006). 'Test-Enhanced Learning: Taking Memory Tests Improves Long-Term Retention', *Psychological Science* 17(3): 249–255.

Rohrer, D. and Taylor, K. (2007). 'The Shuffling of Mathematics Practice Problems Boosts Learning', *Instructional Science* 35: 481–498.

Rosenshine, B., Meister, C. and Chapman, S. (1996). 'Teaching Students to Generate Questions: A Review of the Intervention Studies', *Review of Educational Research* 66: 181–221.

Royal Society (2007). *The UK's Science and Mathematics Teaching Workforce: A 'State of the Nation' Report 2007* (London: Royal Society). Available at: https://royalsociety.org/~/media/Royal_Society_Content/education/policy/state-of-nation/SNR1_full_report.pdf.

Saner, E. (2011). 'How Good is Sex Education in Schools?', *The Guardian* (10 October). Available at: http://www.theguardian.com/lifeandstyle/2011/oct/10/how-good-is-sex-education.

Sebba, J., Crick, R. D., Yu, G., Lawson, H., Harlen, W. and Durant, K. (2008). *Systematic Review of Research Evidence of the Impact on Students in Secondary Schools of Self and Peer Assessment*. Technical Report no. 1614 (London: EPPI-Centre, Social Science Research Unit, Institute of Education, University of London).

Snowball, A., Tachtsidis, I., Popescu, T., Thompson, J., Delazer, M., Zamarian, L., Zhu, T. and Kadosh, R. C. 'Long-Term Enhancement of Brain Function and Cognition Using Cognitive Training and Brain Stimulation', *Current Biology* 23(11) (2013): 987–992.

Strauss, V. (2015). 'The Real Stuff of Schooling: How to Teach Students to Apply Knowledge', *Washington Post* (24 March). Available at: https://www.washingtonpost.com/news/answer-sheet/wp/2015/03/24/the-real-stuff-of-schooling-how-to-teach-students-to-apply-knowledge/.

Sweller, J. (1988). 'Cognitive Load During Problem Solving: Effects on Learning', *Cognitive Science* 12(2): 257–285.

Vaughan, A. (2015). 'Wildlife Thriving Around Chernobyl Nuclear Plant Despite Radiation', *The Guardian* (5 October). Available at: http://www.theguardian.com/environment/2015/oct/05/wildlife-thriving-around-chernobyl-nuclear-plant-despite-radiation.

White, B. and Frederiksen, J. (1998). 'Inquiry, Modelling, and Metacognition: Making Science Accessible to All Students', *Cognition and Instruction* 16(1): 3–118.

Willingham, D. T. (2004). 'Ask the Cognitive Scientist', *American Educator* (summer). Available at: http://www.aft.org/periodical/american-educator/summer-2004/ask-cognitive-scientist.

Willingham, D. T. (2004). 'Practice Makes Perfect – But Only If You Practice Beyond the Point of Perfection', *American Educator* (spring). Available at: http://www.aft.org/periodical/american-educator/spring-2004/ask-cognitive-scientist.

Willingham, D. T. (2010). *Why Don't Students Like School? A Cognitive Scientist Answers Questions About How the Mind Works and What It Means for the Classroom* (San Francisco, CA: Jossey-Bass).

INDEX